Layers of History
Essays on the Khasi-Jaintias

The Author

David R. Syiemlieh joined the Department of History, North-Eastern Hill University (NEHU) in 1979. He was the Head of the Department of History, 2002–2005. He was Honorary Director of the Indian Council of Social Science Research–North Eastern Regional Centre (ICSSR-NERC). A member and for many years Treasurer and Secretary of the North East India History Association, he was nominated in 2005 as Council Member of the Indian Council of Historical Research (ICHR), New Delhi and served two terms as Member of the ICHR. Professor Syiemlieh is former Vice-Chancellor, Rajiv Gandhi University. He has won numerous awards and scholarships such as J.J.M. Nichols-Roy Award; National Scholarship; Jawaharlal Nehru Book Award and NEHU Gold Medal for First Position in MA; Charles Wallace India Trust Grant; India-France Cultural Exchange Programme; and Senior Fulbright Fellowship at Notre Dame University, USA.

Professor Syiemlieh has edited numerous books including volumes of *Proceedings of the North East India History Association* and *On the Edge of Empire: Four British Plans for North East India* (2014). Author of numerous published articles in journals/edited volumes, some of his authored books include *British Administration in Meghalaya Policy and Pattern* (1989), *A Brief History of the Catholic Church in Nagaland* (1990), *They Dared to Hope* (1999) and *Survey of Research in History on North East India 1970–1990* (2000). Professor Syiemlieh is presently Member, Union Public Service Commission.

Layers of History
Essays on the Khasi-Jaintias

David R. Syiemlieh

2015

Regency Publications

A Division of

Astral International Pvt. Ltd.

New Delhi – 110 002

Cataloging in Publication Data–DK
Courtesy: D.K. Agencies (P) Ltd. <docinfo@dkagencies.com>
Syiemlieh, David R. (David Reid), author.
Layers of history : essays on the Khasi-Jaintias / David R. Syiemlieh.
    pages   cm
Includes bibliographical references.
ISBN 9789351307433 (International Edition)

    1. Khasi (Indic people)–India, Northeastern–History. 2. Jaintia (Indic people)–India, Northeastern–History. 3. India, Northeastern–History.   I. Title.
    DDC 305.8009541   23

Published by   : **Regency Publications**
                         A Division of
                         Astral International Pvt. Ltd.
                         – ISO 9001:2008 Certified Company –
                         4760-61/23, Ansari Road, Darya Ganj
                         New Delhi-110 002
                         Ph. 011-43549197, 23278134
                         E-mail: info@astralint.com
                         Website: www.astralint.com

Laser Typesetting   : **Classic Computer Services**, Delhi - 110 035

Printed at   : **Thomson Press India Limited**

PRINTED IN INDIA

For

Professors J.B. Bhattacharjee, Imdad Hussain, Jai Prakash Singh and Milton Sangma

Teachers, Colleagues and Friends

Acknowledgements

I am grateful to many persons and institutions for their support. Permission was kindly provided by the Department of History, NEHU to publish three articles in the SAP series on *Economy and Society in North East India*. The North East India History Association, kindly gave permission to publish articles originally placed in several Proceedings of the North East India History Association. The Vivekananda Kendra Institute of Culture, Guwahati is gratefully acknowledged for permission to publish the essay indicted. The Indian Council of Social Science Research, North Eastern Regional Centre, Shillong, has given permission to reprint the essay on traditional institutions. The Indian Council of Historical Research, New Delhi, has given permission to reprint the essay on sources in archives and libraries in the UK. Prof. J. B. Bhattacharjee of the Institute of Northeast India Studies, Kolkata, has given permission to incorporate the essay on Homiwell Lyngdoh. The Centre for North East Studies and Policy Research , Jamia Millia Islamia, New Delhi, has consented to place the essay on Tirot Sing in this volume. Dr. O. L. Snaitang and Malcolm David Roy have given permission to incorporate the essays on Hajom Kissor Singh and David Roy. *The Shillong Times* has given permission to reprint the essay on Thomas Jones.

I have been fortunate to have researched in several archives and libraries in the pursuit of historical material for this collection and other interests. I acknowledge the support given to me by the Librarians, Archivists and Staff of the Assam State Archives, Dispur; the West Bengal State Archives, Kolkata; the National Library of India, Kolkata; the National Archives of India, New Delhi; India Office and Records now integrated into the British Library, London; Nottingham University Archives and the North-Eastern Hill University Library.

Preparing the text for editing required the expertise of Mr. Godfrey Pathaw of NEHU and Mr. Rengarajan at the UPSC, both of whose support is gratefully acknowledged. I am grateful to Mr. Anil Mittal of Regency/ Astral Publications for publishing this collection of essays on the Khasi- Jaintias.

I am indebted to NEHU for the opportunity the University gave me to teach and research. My students and scholars shared my excitement researching on North East India. I remember them dearly. The institutions and persons mentioned are acknowledged for their kindness and support. My family has been most encouraging. To all I say 'Khublei'.

David R. Syiemlieh

Preface

Historical research on the history of North East India was in part encouraged by the administrative reorganization of the region and the establishment of several universities and research organisations. Continuing the tradition of research initiated by Gauhati University with its broad perspective of studies on the region, the Department of History, North-Eastern Hill University, in which I was a faculty for many years, expanded the enquiries both in theme and perspective. NEHU enabled scholars and teachers to work on variety of histories in an encouraging academic environment. Encouraged by senior faculty who shared their experience and knowledge the Department has over the years been able to build up a corpus of literature on many aspects and facets of the histories of the people. In large part this was possible though the liberal funding of the Special Assistance Programme provided by the UGC.

All but one of the essays in this collection was researched and published during my NEHU years. All the essays have been looked at afresh and updated. The first publication/lecture of an essay is indicated in the endnotes.

The essays relate to the Khasi-Jaintias and cover a time span from their pre-colonial past, through the colonial era and into more contemporary times. The essays do not purport to be a complete modern history of the tribe-there are several areas of their history that have not been covered in this collection as they have been studied by other colleagues and scholars. *Layers of History* begins with an essay on the division of history of the tribe into time phases and critiques the applicability of these divisions. An essay on the research material on the Khasi- Jaintias in libraries and archives in the UK follows. Several essays in the collection relate to the trade and economic activity of the Khasi-Jaintias. Four essays turn attention to the Khasi states - British policy toward the states, *Syiemship* succession, the Federation of the Khasis states and the incorporation of states into the Khasi-Jaintia Hills district and India. Of the many more significant and silent makers of its history several stand out prominently. The collection includes essays on David Scott, Tirot Sing, Thomas Jones, Hajom Kissor Singh and David Roy. A rather long essay takes the discussion to the traditional governance of the Khasi hills and brings it up to more recent times. The last essay provides a perspective of the people, the land and the State.

My four colleagues mentioned in the dedication have been of immense support to me. My skills as a historian were first tuned by Professors Jayanta B. Bhattacharjee and Imdad Hussain. From both these colleagues I learned the expertise of searching for, interpreting and analysing the material collected. J. B. Bhattacharjee's understanding of the history of the region is phenomenal - it spans prehistory to contemporary times. He has written most extensively on the region. Beginning with his early research on the Garos and later Cachar, his studies on state formation and trade have been well received. An able administrator and visionary, the North East India History Association continues to be his passion. Professor Imdad Hussain is known for his fine classes and lectures and keynote addresses. His readings on the history of the region and beyond it frontiers with Tibet and Burma is remarkable. His history of Raj Bhavan, Shillong is a much sought after book. I have benefited immensely from their teaching, supervision, encouragement and friendship. We have had many a discussion on the history of our interests, including several of the essays included in this collection.

Professor J. P. Singh is a renowned numismatist. His amiable disposition has left a fine imprint on many he came in contact with. Professor Milton Sangma's contribution to the history of the region is his *History and Culture of the Garos* and the history the American Baptist Mission and Christianity in North East India. He continues to write and lecture in his native Garo.

I am fortunate to have had them as teachers, colleagues and friends.

David R. Syiemlieh

Contents

Acknowledgements   *vii*

Preface   *ix*

1. Periodizing North East India History   1

2. Sources on the History of the Khasi-Jaintias in Libraries
    and Archives in the United Kingdom   7

3. Technology and Socio-Economic Linkages of the Khasi-Jaintias
    in Pre-Colonial Times   17

4. Control of the Foothills: Khasi-Jaintia Trade and Markets
    in the Late Eighteenth Century   29

5. Trade and Markets in the Khasi Jaintia Hills: Changed
    Conditions in the 19th and 20th Centuries   37

6. Colonialism and Syiemship Succession: A Study of Cherra State
    (1901-1902)   47

7. British Policy Towards the Khasi States   55

8. The Federation of Khasi States: Three Phases of its History   65

9. The Integration of the Khasi States into the Indian Union   71

10. Call of Freedom from the Hills: Tirot Sing and his Significance
    in the Freedom Struggle   79

11. The Last Days of David Scott   87

12. Remembering Thomas Jones   95

13. Among Many Writers: Contribution of Homiwell Lyngdoh
    to Khasi History   101

14. Indigenous Roots: Hajom Kissor Singh and the Founding of
    Unitarianism in the Khasi-Jaintia Hills   109

15. David Roy: Notes on the Khasis   115

16. Traditional Institutions of Governance in the Hills of
    North East India: The Khasi Experience   125

17. The North East: Home of many Tribes and Communities   139

1

Periodizing North East India History

Introduction

Indian history, despite its own historical traditions was drawn to influences from the west and by western writers. From the early part of the twentieth century when Indian history began to be taken as an academic subject, its understanding of periodization was drawn largely on European models.[1] The Positivist approach drew inspiration from the Marxian tradition, and the Cambridge and imperialist schools from the university after that name. The Nationalist tradition was drawn from the realization by Indian nationalist, many of whom had studied in British universities, that it was their responsibility to provide an account in the understanding of India's past. There is justified reason for objecting to the classical if not archaic subdivision of Indian history into 'Ancient', 'Medieval' and 'British' periods. If somehow we accept the nomenclature 'Ancient' and 'Medieval' as they are based on variable time scales, there was just no credible reason to have a 'British' period in Indian history as it negates a history of the people of India. Within these broad phases, history tended to be written with the dynastic reigns of the Guptas, Mauryas, Tughlaqs,and the Mughals and British rule and going within this broad frame into the dynastic histories of the Palas, Senas and coming closer home the Ahom period.

Though there has been a continuous history of India going into millennia, the periodization of Indian history was more influenced on political lines and not on the social, economic and cultural institutions.[2] Historians have to see the significant in change to arrive at the close of an era and the beginning of another. Historians have tried to classify history and to periodize history on the basis of some major event. The close of the ancient early India could be dated to the Arab conquest of Sind (AD 712) or AD 1000, even pushing the date to the establishment of the Delhi Sultanate (AD 1226). While at this, it has become more acceptable to use the term 'Early India' for 'Ancient' and 'Sultanate' or 'Mughal' for 'Medieval' largely after the politically significant developments.

While some historians prefer 1757 as a convenient date for ushering the 'Modern' or 'Colonial' phase, other prefer the date AD 1765 for reasons of the impact the signing of the Diwani of Bengal had for the subcontinent. Even if it is desirable to fit history into time scales the question arises where and when should this phase begin? Where do you end? If the 'modern' has far more distinctive features in political, economic and social features including colonialism in its many forms, the nationalist movement and India's independence and after, and not to be too insular, all this within the global developments, another problem arises. It is that the 'modern' history is getting longer and longer. Just as the starting point of this period is debatable, there is no consensus on the close of this phase of Indian history. Romila Thapar's division of Indian History into twelve periods allotting nine to 'early' Indian history, two to 'medieval' and only one to "British colonial rule and the Indian nationalist response" raises questions of what length in time each period may be provided and why there should be such load on the 'early' to the neglect of 'medieval' and 'modern'.[3] If 'early' history spreads over a much longer time span, 'medieval' is squeezed in between this phase and 'modern,' the time frame of 'modern' is seen as having sufficient elasticity to move into 'contemporary.' Even as this vexed question of periodization continues, there is a continued search for understanding Indian history on European parallels as has recently been attempted by Catherine Asher and Cynthia Talbot.[4]

Periodizarion of Indian history does not necessarily require conforming to the dates of world and European histories. Ideally each history should be dated by their own locale and the historical significance. And yet history incorporates the local into the general. Indian history is replete with examples of this situation. I will close this section on the problems of periodization with a thought that will require debate and discussion. Historically many tribes and communities of the North East were not really part of the Indian ethos till more recent times. Historians by giving them an 'ancient/ early', 'medieval' or 'modern' connection and past within the Indian context and without the people's consent, require explaining their position in this regard.

North East India

Something now requires to be said of the periodization of history with particular reference to the hill areas of North East India. Indian history has generally been fitted into one or more divisions of 'pre-history', 'ancient', 'medieval', 'modern' and 'contemporary'. Alternately there have been studies that have not used the time schedule rather the concepts such as 'pre-colonial', 'colonial' and 'post-colonial'. Both the patterns referred to have been applied to the Brahmaputra, Barak and Manipur valleys because all the four divisions / concepts provide a continuum of history in these areas. This is because these areas have material remains and written material for reconstructing their history. In the case of the hill areas of the North East such periodizations are not applicable for a number of reasons. While it may be argued that it is possible to reconstruct the pre-history of some of the tribes from material remains, it becomes difficult to provide the tribal histories with 'ancient' and 'medieval' pasts because there is an absence of written material from which to reconstruct that past. Even giving them a modern history becomes difficult, because to move into the modern without an intelligent account of the two earlier periods of history would be contributing to a fallacy in history.[5] The difficulties in such a periodization may be overcome. Amalendu Guha a former President of NEIHA suggested in an address at a seminar at the North- Eastern Hill University that:[6]

> In the case of most parts of North East India, periodization breaks down since major segments of our population have lived down to the early 19th century without mastering the art of writing. If we want to know about their pre-19th century past, then the conventional methods of writing history do not help. We need to take resort to research in linguistics, ethnology and archeology for the purpose.

Another difficulty in applying the term 'pre-colonial' to several of the tribes in the region is that the concept is often understood to cover their history over several centuries after their pre-history.[7] There are references to several of the tribes in the 'ancient' Puranic texts and in later literature over a rather long span of time. There are references to them in the 'medieval' Ahom Buranjis, coins, and inscriptions, apart from references in Bengali texts and early 'modern' European-English reports. While three phases of periodization have been applied to the general histories of India, for the hill tribes of the region, and perhaps elsewhere too, where the concept 'pre-colonial' has been used it may be prudent, where material is available and can be dated with certainty, to add the words 'early', 'later' as the case may be.

In as much as the terms 'ancient', 'medieval' and 'modern' are not quite applicable to interpret the histories of the hill tribes, the term 'pre-colonial' would vary in time from tribe to tribe. In the case of the Jaintias for instance, this phase of their past would cover many centuries, possibly through the 13th-18th centuries. A general reference to Naga history may draw references to these same dates. However, when the histories of individual Naga tribes are attempted and particularly to those, which did not have an interface with the people of the Assam and Imphal valleys, these concepts in time become difficult to apply. Such would also be the case with the tribes of Arunachal Pradesh and Mizoram and several other smaller communities in the region. It has been found difficult therefore to assign dates with certainty to the histories of the tribes before historical material becomes available to reconstruct their past. One way out of this dilemma will be to use, where available, as has been suggested, literature in other languages on the tribe under study. In the case of the Khasi- Jaintias the use of Bengali and Assamese literature would help get across the problem of sources. It would be interesting, as it has yet not been done, to study Mizo history before the British connection using Burmese and Bengali sources in addition to the oral traditions of the tribe. Similarly Burmese, Tibetan and Assamese sources would be useful to reconstruct something of early Naga history and that of several of the Arunachal tribes.

Whatever is the debate, people and communities will recollect their past and have their histories written. In a nation of such diversity as India, no general periodizing scheme will work for all groups and in particular for the communities of North East India. Let all histories be written with prime concern for the past in the *Itihasa* tradition "so indeed it was".[8] Meanwhile let the debate on periodization continue, for periodization will not alter history.

Notes and References

This first essay is an extract from David R. Syiemlieh, Presidential address , 'Periodization: Mission History,' at the Annual Session of the North East India History Association, Don Bosco College, Tura, 10 February 2011.

1. All three major schools of Indian historiography have European influence. Among other literature on this subject read Romila Thapar, 'Ideology and Interpretation of Early Indian History', *History and Beyond*, Oxford University Press, New Delhi, Fourth Impression, 2006, pp.1-22; and Romila Thapar 'Perceptions of the Past.,' *The Penguin History of Early India from the Origins to AD 1300"*, Penguin Books India, New Delhi, 2003, pp.1-29.

2. Partha Chatterjee, 'History and the Domain of the Popular' ,www . indiaseminar.com /.../522%20partha %2 0 chatterjee . htm.

3. R. Thapar, *The Penguin History of Early India from the Origins to AD 1300* , pp.29-32.

4. A recent review of Catherine B. Asher and Cynthia Talbot's, *India Before Europe*, Cambridge University Press, 2006, mentions: "Over the past several decades, the way historians have divided up India's past has changed and been brought more in line with the periodization of European history. Medieval, now refers to the centuries from either 500 or 700 CE up to approximately 1500, while the 300 years from 1500 and 1800 are called early modern, just as in the case of Europe. Although the new scheme of periodization is easier to grasp due to its similarity to the way Europe's history is conceptualized, this is not the primary reason it was adopted. There are good justifications internal to Indian history why the 1,000 years between 500 and 1500 should be considered as a single unit of the past. Shortly after 500, the last of the ancient Indian empires disappeared and for the next millennium India's states were typically smaller and more decentralized. Between approximately 1500 and 1800, as we argue in chapter 6, larger and more efficient states once again emerge in India, along with a host of other phenomena, such as commercialization, that are characteristic of the early modern period throughout much of the world. Since the central concerns of many historians are political structures and the economic systems that underpin them, the new periodization of Indian history with a dividing point at 1500 is more appropriate".

5. David R. Syiemlieh, 'History Writing on North East India: Periodization, Varieties, Concerns,' Bharati Ray (ed.), D. P. Chattapadhyay (General Editor), *History of Science, Philosophy, and Culture*, vol. XIV, Part 4, *Different Types of History*, Pearson Longmans, New Delhi, 2009, p.234. Hamlet Bareh in *The History and Culture of the Khasi People*, Shillong , 1985, has divided Khasi history into pre-history and early history, medieval and pre-British period and modern history. Similarly B.B. Gupta in *History of Nagaland*, New Delhi, 1982, has divided Naga history into ancient, medieval, modern and cultural, and Milton Sangma in *History and Culture of the Garos*, New Delhi, 1981, has sections on the ancient and medieval histories. Their attempt to provide comprehensive histories for the Khasis, Nagas and Garos and the scanty material on which to reconstruct the past before the 19th century is reflected in the chapters before the more detailed modern histories.

6.   Amalendu Guha, 'Introduction', in Gautam Sengupta and J. P. Singh, (eds.) *Archaeology of North Eastern India,* New Delhi, 1991,p. 2.

7.   Read Manorama Sharma's comments on this problem in 'Socio-Economic History in Pre-Colonial North East India: Trends, Problems and Possibilities', Mignonette Momin and Cecile A. Mawlong (eds.), *Society and Economy in North-East India*, vol. 1, Regency Publications, New Delhi, 2004, pp.1-2; pp.16-17;   Manorama Sharma, 'Trends, Possibilities and Problems in History Writing in Pre-Colonial North-East India: Economy and Society' , Fozail Ahmad Qadri (ed.), *Society and Economy in North-East India,* Vol. 2, Regency Publications, New Delhi, 2006, pp.16-17;   David R. Syiemlieh, 'Technology and Socio-Economic Linkages of the Khasi-Jaintia in Pre-Colonial Times',   M. Momin and Cecile Mawlong (eds.), *Society and Economy in North East India,* vol. I , Regency Publications, New Delhi, 2004.pp.22-23;   and David R. Syiemlieh, 'Control of the Foothills: Khasi Jaintia Trade and Markets in the late Eighteenth Century', F. A. Qadri (ed.), *Society and Economy in North-East India*, Vol. 2, Regency Publications ,New Delhi, 2006 , p.327.

8.   B. N. Mukherjee, 'Reflections on Trends in Indian Historiography', *Presidential Address, Indian History Congress,* Sixty-First Session, Calcutta, 2001, p.10.

2

Sources on the History of the Khasi-Jaintias in Libraries and Archives in the United Kingdom

Introduction

The Khasi-Jaintia connection with the British goes back to the late 18th century. The East India Company's rule over Bengal brought it into restrained control over the Khasi-Jaintias living in the foot-hill of the plateau adjoining Sylhet. Company's rule was not extended over the hill people till the third decade of the 19th century. Their initial interest in trade and commerce later broadened into political interference, which by 1830-1835 was followed in the annexation of the Jaintia hills into the British dominion and a political control over the twenty-five Khasi states. Other forms of interaction were to come in the wake of the opening of hills with the interest of a number of mission societies in spreading Christianity among these hill people. Meanwhile, the Khasi resistance to British rule being suppressed by 1833-1834, experiments were conducted by the Company to establish a sanatorium and a cantonment for European troops in these hills. By early 1834, the experiment had shown a lack of interest in Cherrapunji in preference for Darjeeling, but the station's importance continued with it becoming the headquarters of the Khasi Hill Political Agency in 1835.

While the Khasis apparently had come to accept their fate, either directly under British rule, as in the case of 31 British villages, or indirectly in the case of the 25

Khasi states, the Jaintias, having little semblance of British authority over them, reacted to foreign rule. Over two phases in resistance between 1860-1863, the Jaintias were engaged in the 'little war' of North East India as an expression of their resentment against British rule over their hills. The later decades of the 19th century unto the end of colonial rule, however, witnessed a comfortable adjustment of little interference by Government and least resistance by the community.

Location, Content and Use of Sources

Much of the material used in reconstructing the history of the Khasi-Jaintias over the period of their interaction with the British have hitherto been drawn from local sources in the form of oral traditions and a variety of written materials in Bengali, Khasi and English located in libraries and archives at Shillong, Guwahati, and New Delhi. Few historians have so far been able to use the impressive collection and variety of records, memories, private correspondence and accounts left by British administrators, soldiers, botanists, missionaries, traders and even the casual visitors to the Khasi-Jaintia hills that are available in libraries and archives in the United Kingdom. The purpose of this paper, therefore, is to provide information on the location of the material in the United Kingdom, their content and the use of the material for reconstructing the history of the Khasi-Jaintias.

London

The India Office Library and Records, now the Oriental and India Office Collection in the British Library at St. Pancras, London, has the largest collection of manuscripts, books and records on India. Its records comprise the archives of the East India Company (1600-1858), the Board of Control (1784-1858), and the India Office (1858-1947). These voluminous records have numerous references to the Khasi-Jaintias. Some of these records are linked with and are available for consultation in the West Bengal Archives, Kolkata, and the National Archives, New Delhi. For a more complete understanding of the operation of British rule, its administration and policy decisions, the consultation of records in the Indian Office is invaluable.

The private collections in this Library include many references to the Khasi-Jaintia hills and its people. Those of John Colpoys Haughton (1817-1887) who was Civil Commissioner in these hills when the Jaintia resistance raged, has a number of correspondences between the Commissioner and others on the suppression of the disturbances. The collection comprises the Letters Books of J. C. Haughton in two volumes (MSS. Eur. D. 529) and the letters of J.C Haughton (MSS. Eur.D.530). Book 1 of the Letter Books includes his correspondence with Lionel Inglis whom he advised 'to be vigilant to prevent strong parties of plunderers from going down towards Assam'.[1] A number of letters are also addressed to B.W.D. Morton, the district's Deputy Commissioner, In one of these he advised that 'when any rebel is caught and brought before you, you should at once regularly try and sentence him, the substance of all the proceedings being recorded in your own hand.'[2] This letter continues: 'I would say the traitors caught may well be hanged but surrendering persons, unless guilty of actual murder, might be, guaranteed their lives but no more'. Haughton regularly kept Cecil Beadon, the Lt. Governor of Bengal in Calcutta informed of the developments in the hills and gave suggestions for a speedy return to peace.[3] A letter is addressed to Major Thelwall, in-charge of a section of the British forces engaged in the Jaintia campaign.[4]

The letters of J.C. Haughton are all private correspondences with his father. There is an interesting letter he writes in which he mentions that he had the satisfaction of sentencing the very last of the rebels to transportation. "He was a most desperate ruffian," he wrote, "and it is believed had committed many murders in cold blood. Yet I find one of the papers actually censoring me for not pardoning him!" Of the Jaintias who were tried, Haughton informs that three were sentenced to transportation, two to death and about twenty to short periods of imprisonment by the Military Commissioners with special powers appointed by the Government, "and with whom I could not interfere if I desired to do so."[5]

Another large collection with the Oriental and India Office Collection are the papers of Robert N. Reid Governor of Assam between 1937-1942. His diaries, notes, letters, photographs and collection of articles covering two boxes of many volumes were deposited with the Library by Lady Reid and Mr. E.D. Reid in 1965 and 1980. A very informative source is his unpublished manuscript "Assam and the North East Frontier of India."[6] Written in 1946 when he was home on retirement the manuscript has many references to the Khasi and Jaintia hills, their administrative and constitutional positions and the plans he had for these and other hill areas in India's North East. In his collection is also a copy of a document he wrote while in Assam. 'A Note on the Future of the Present Excluded and Partially Excluded and Tribal areas of Assam',[7] as the title suggests was meant to examine the position of these three areas in the last years of British rule. Researchers may also draw useful material from his speeches[8] and articles that he wrote following retirement and return home.

Still other collections that have some references to the Khasi-Jaintias are the Charles Wood Collection, (MSS Bur. F. 78/89/Bundle 3), which has something on the aftermath of the Jaintia resistance,[9] the J.C. Hutton papers (microfilm 2406 B) on the reforms proposed in the 1930s preceding the Government of India Act 1935; the Linlithgow collection, (MSS. Eur. F./125) that has correspondence between the Governor and the Governor-General; and the Political Department Collection, (particularly files L/P and J / 6787, 6789 and 10635) on developments in the North East hills just prior to the transfer of power and after. The ardent researcher will still find much more material in this great storehouse of information. The India Office Library and Records guide *Select List of Private Collections in the European Manuscripts* only gives an indication of some of the more important collections held in the archives.

Nottingham

The private and some official letters and correspondence of Lord William Bentinck are deposited in the University of Nottingham, Manuscript Department. These records and letters are a useful source for studying the early years of British control over Assam and the Khasi Hills.[10] The Bentinck papers form part of the Portland Collection deposited by the Duke of Portland with the University in 1949. While Bentinck's personal correspondence on the Cherrapunji experiments are few, the letters he received from George Swinton, the Chief Secretary of the Bengal Government; R.

Benson, his Military Secretary; David Scott, Agent to the Governor-General, North East Frontier, and military, medical and other officers involved in the Cherrapunji sanatorium, are impressive in terms of information they provide in size and content.

The collections relevant to our study has been catalogued as follows:

PWJF 2781 (1829-1833) Letters and Enclosures from George Swinton.

PWYF 2791 (1830) Correspondence concerned with North East Frontier.

PWJF 2811 (1831) Correspondence with George Swinton.

PWJF 2820 (1831) Correspondence with George Swinton.

PWJF 2857 (1832-1833) Miscellaneous (including letters) from George Swinton.

Surprisingly, these papers do not provide much information on the Khasi Uprising of 1829-1833 other than a reference or two of the 'bow and arrow insurrection".[11] Nor is there anything on the developments prior to the incorporation of the Jaintia Raj into the British dominion. The collection, however, has to offer much material on the official view of the British connection with Khasi hills from 1829 to 1834. The private letters also make very interesting reading, particularly the concern David Scott's friends had for his health, details of his death,[12] Scott's own interest in the economic development of the Khasi hills, the construction of the road, and survey operations by Capt. Fisher, the discovery of coal in Cherrapunji and the plans to set up a cantonment at Cherrapunji and the beginning of problems and the eventual collapse of the Cherrapunji cantonment plans in 1834.[13]

Edinburgh

The Nottingham University muniments come to significance when read with the material found in the papers of the 9th Earl of Dalhousie on deposit in the Scottish Records Office, Edinburgh. The Earl, father of the more renowned Governor-General, was Commander-in-Chief in India between 1829-1832 and the papers listed in GD 45/5 relate to this service. An interesting deposit is GD/ 45/48. It is a letter from George Swinton at Calcutta with two enclosures, one a map of the "Cossya independent estates" and the other an extract of a letter from Scott to Swinton dated 27 October 1830. The map referred to was the first to be made of the Khasi states and is dated 1829. It illustrates and delineates the territories of some of the large states that the British had come to know. It indicates the line of the road that was to be built between Jaintiapur and Raha. Scott's letter to George Swinton, his Fort William College friend and then Chief Secretary, Government of India, throw light on a variety of subjects such as the life of European troops stationed in India, the advantages of Cherrapunji, the drinking habits of soldiers and more. He wrote.[14]

I am more than ever convinced of the immense advantage that the European force would derive from being cantoned in this quarter. If the recruits were at once brought up here they have no opportunity of acquiring idle, dissolute habits and they would remain vastly superior as soldiers.

Scott had other worries of the men stationed at Cherrapunji. He informed Swinton in his own hand:[15]

They have all behaved exceedingly well the Cossyas being ready in all other matters to be hale fellows well met and having no objection to give any of them a glass of grog in the very penetralia of their houses. The carpenters, platers, etc., are, all men that cannot be trusted out of sight and I do not therefore like taking any of them over to Myrong.

Other materials useful for the history of early British administration in Eastern India are dealt within the administration of justice in the Non-Regulation Provinces (GO 45/6/137); papers relating to the Ghoorka Corps (GO 45/5/26); Memorandum on the relations between civil and military authorities (GO 45/5/50); Letters and copies of letters to and from Lord William Bentinck (GD 45/5/60); and finally, Bengal Army List (1831), (GO 45/5/84).

Many British families had tradition of keeping their members private and official correspondence, notes, memoirs as the Bentinck and Dalhousie records indicate. For their safe custody and restricted use, many of these eventually have found their way into Universities, libraries and archives, though some continue to be in smaller private holdings. Another collection that has become useful for studies on Eastern Bengal, particularly Sylhet and Cachar and the Khasi and Jaintia hills, is the Robert Lindsay Muniments. Lindsay arrived in Calcutta in the autumn of 1772 as an eighteen-year-old writer. After occupying various subordinate posts under the Calcutta Presidency, he was transferred on promotion to Dacca in 1776. In August, the following year, he was appointed as Superintendent of the frontier district of Sylhet, He remained there as its Collector for twelve years. Apart from his official duties, which ranged from revenue collection, judicial and executive administration, Lindsay, like 'Sylhet' Thackeray before him, engaged himself in a variety of trading activities ranging from limestone (*chunam*) trade with the Khasis, elephant hunting, construction of ships, grain export, collection of cowries on a commission, and sale of jute from lower Assam to Bengal. From a combination of all those enterprises he amassed a considerable fortune and enough to see him through early retirement. In 1822 he wrote his biography "Mr. Robert Lindsay: Memoir of His Life".

Lindsay kept a meticulous account of all his private and official correspondence. These muniments were brought back with Lindsay following his resignation. These forms part of the much larger Crawford and Balcarres papers. In 1946, the entire collection was transferred from the family archive to the John Rylands University Library of Manchester. There they remained till 1988 when they were shifted to the National Library of Scotland in Edinburgh.

The papers of Robert Lindsay are listed: 9769/30/5. Of the manuscript volumes those with numbers 9769/30/5/1 to 9 and 18 relate to his Indian experience. Volumes 10 to 20, excluding 18, give an account of his business transactions, and ledgers and correspondence on his return home till his death in 1836. Indian Baggage Book 1784 is a slim volume which lists the baggage Lindsay sent home, possibly on one of his own ships. What is of interest to us are the contents he lists in Box 4, 5 and 6. Here were packed 2 hill dresses containing five pieces each; 3 ivory quivers mounted on brass and silver; bundles of arrows; a rhinoceros head and 3 horns; a tiger's head, swords, bows and hill shields. Lindsay presumably had made frequent visits into the hills. Indeed, he just might have been the first European to have gone beyond the limestone foothills.[16]

The most important volumes for information on Lindsay's commercial activities are 9769/30/5/4, 5 and 9. Here he writes in detail of how he got interested in the limestone trade, the names of his employees, the patronage he gave to William Raitt, his brother Henry and other European and Greek traders who were taking interest in the *chunam* and salt trade. 9769/30/5/8 are the three volumes John Shaw and George Reid: Estate Accounts 1782-92'. In volume II he notes his worry that the defense of Sylhet was insufficient. [17]

> From its vicinity to the mountains we are surrounded on three sides by troublesome neighbours. Acting offensively against them we have found from long experience ineffectual and the only mode of ensuring their good behaviour and of preventing their incursions is by supporting armed boats upon the lakes and several forts and passes leading towards the hills.

Lindsay found that 'the hill people are most troublesome (at) the beginning of harvest.' In a letter to Lt. James Davidson, he suggested him to 'prevent the Hill people from cutting crop and carrying the rice to the mountains'. Ten years' experience had taught him that there was, "no other mode of preserving their peaceable behaviour and of enforcing them to our accommodation than by stopping their usual supply of grain -this I have always found had the desired effect in the course of two or three months. [18]

In volume II of Linday's 'India Letter Books' is to be found, possibly, the first report of information on the Khasis written in the English language. [19] Lindsay was then in Calcutta and had tendered his resignation as Superintendent of Sylhet. The Board of Revenue's letter to the Scotsman informed him that his resignation had been accepted and directed him to furnish a particular description of the life of the Cosseahs (Khasis), the nature of the country they inhabit and their particular habit of life. Lindsay promptly replied for he was eager to sail home. This very short account [20] has no details of the Khasis, but some insights are there. For instance, he refers to their eating habits; the cremation of their dead; the independence of each of the small Khasi states and the reaction to those who do them wrong. Even two hundred years ago this observer could say that the Khasi women were industrious having observed them bring down from the hills iron to the markets in Sylhet and their activities in the border *hats*.

The only part of Lindsay's papers that have been read, other than the reference mentioned above, is his memoir. [21] Written in 1822 and entitled "Mr Robert Lindsay: Memoir of His Life", the biography was published with a new title "Anecdotes of an Indian Life" and placed in the collection *Lives of the Lindsays or A Memoir of the House of Crawford and Balcarres* (vol. III, London, 1849). The 77 pages memoir has become an important source on the early British trade with the Khasis; it has references to the trade in iron and jute; has descriptions of the Khasi men and women and their habits. Lindsay recollects a hunting expedition he made into the hills accompanied by the Jaintia Raja. [22] Lindsay writes much on Sylhet where he served and traded for many years.

British Museum

The material located and used and referred-to above could not have been possible without M.D. Wainwright and Noel Mathews' authoritative guide as a starting point.[23] The editors have produced a very useful reference after an extensive study of the nature and contents of the 'private' sources.[24] One very interesting manuscript was followed from a reference in the guide. Lt. Edward H. Steel's "Account of the Khasiah Tribe on the Borders of Eastern Bengal and Assam" in the "Nightingale Collection"[25] is located in the Western Manuscripts section of the British Library in the British Museum. The 24-page account (and there is mention of photographs which just might be the first images of Khasi- Jaintias), sketches and artifacts were purchased by the Museum from Major Golding and Lawrence in January 1877.

Lt. Steel of the Royal Artillery was stationed in the Khasi-Jaintia hills long enough for him to write the account with some academic interest. In all probability he may have seen service in the Jaintia campaign and the transfer of the district headquarters from Cherrapunji to Shillong. The manuscript starts with a description, in the first few pages, of the Khasi hills, the neighbouring plains areas, the climate and so on. From page 5 onwards, he describes the Khasi physique, their dress, the status of their women, their customs at birth, marriage and death, Khasi dances, their basketry, the erection of megaliths and their economic activities, including iron smelting, agriculture, their defense mechanism and finally, fishing.[26] Though Steel does not go into any detail, his account is useful for the observations he makes. The account has not been referred to by scholars. Its more extensive use of it promises to open up hitherto unknown details of the Khasis of that period in time. Indeed if his photographs are traced, they could be the first photographs of the Khasis. He writes after witnessing a Khasi dance, 'The demure looks of the girls, some pretty enough, and the ardent glances of the youth as they pass round and peep shyly at their lovers is amusing enough and makes a pretty picture.'[27] This picture, together with those of the monoliths and his Khasi servant need to be located and preserved for their value in capturing time in photographs.

Mission Records

The material on the Welsh connection with the Khasi-Jaintias has been utilized by missionaries in their histories of the mission.[28] However, the histories in the Welsh and English languages published in the U.K. are largely drawn from material available there, and those published recently in India on the Welsh Mission in the Khasi-Jaintia hills has been drawn from sources in this mission field. Nigel Jenkins and Andrew May have used wider choice of materials including records in Welsh. In an article in a popular Welsh journal[29] and more recently a book[30] he has indicated that there is so much more to locate and use. Nigel Jenkins has translated with notes Welsh into English; a number of books and reports on this connection from what are available in Swansea and Aberystwyth. Their use in the church histories would be enriched when used-judiciously. Andrew May's recent publications have drawn extensively from archival sources of the Welsh Presbyterian and London Missionary Society archives.[31]

There is a *raj* nostalgia. A smaller version of this may be called the 'mission' nostalgia. Missionaries who served in these hills wrote letters home and told of their experiences. This genre of correspondences has become useful to understand the missionary mind and work. Alun Irfonwy Bannister's translation from Welsh into English of letters of his great-grandfather, Rev. William Pryse, among the earliest of the Welsh Missionaries to have served in the Khasi hills and Sylhet (1849-1866) has become a significant text.[32] The foreword, 2 maps, 6 letters, extracts from two articles in Welsh make the 48 page collection delightful reading and adds significantly as a source for the study of the Welsh Presbyterian mission among the Khasi-Jaintias. No study of the Welsh Presbyterian Mission in these hills will be complete without reference to the records in Welsh and English in Aberystwyth, Bangor, and Cardiff and possibly in many private collections across Wales.

The Cambridge University South Asia Archive is another fine repository which has not been sufficiently gone through for material on the Khasis. Even a cursory glance through its Principal Collection of Papers, edited by Lionel Carter and Dusha Bateson indicate a variety of material deposited in the Archives. Similarly, the Public Records Office and the Royal Botanical Gardens; both in Kew, Richmond would have sources not yet utilised. In the Botanical Garden's library are the diary and notes of the botanist, William Griffiths who stayed for some months in the Khasi-Jaintia hills in the mid 1830s.[33] Useful material may also be with the London Missionary Society whose missionary Jacob Tomlin[34] came to the Khasi hills in 1837-1838.

The source material in the libraries and archives in the United Kingdom mentioned above, and there could be much more, is an indication of their volume, significance and usefulness. They are generally well-preserved and easily available for research. The British past with India and these hills in particular could have different response but as one Welshman wrote: 'the past need not be passed'.[35]

Notes and References

"Sources of the History of the Khasi-Jaintias in Libraries and Archives in the United Kingdom", Ranju Bezbaruah (ed.), *North East India: Interpreting the Sources of its History,* Indian Council of Historical Research, New Delhi, 2006, pp.198-206.

1.   Letter dated Cherrapunji, 10 November 1862

2.   Letters dated Cherrapunji, 11 November 1862; 18 December 1862; 19 December 1862; 14 January 1863, and Maflong (Mawphlang), 24 April 1863.

3.   Letters dated Cherrapunji, 14 November 1862; 8 January 1863; 15 April 1863.

4.   Letter dated Cherrapunji, 11 January 1863.

5.   MSS. Eur. D. 530/214, letter dated 31 January 1864.

6.   MSS. Eur. E. 278/19.

7.   MSS. Eur. E. 278/4.

8.   MSS. Eur. E. 278/11.

9.   Cited in D.R. Syiemlieh, *British Administration in Meghalaya: Policy and Pattern*, Heritage Publishers, New Delhi, 1989, p. 95.

10. *Ibid.*, and N.K. Barooah, *David Scott in North-East India*, New Delhi, 1970,

11. C.H. Phillips (ed.), *The Correspondence of Lord William Cavendish Bentinck*, vol. I, Oxford, 1977, No. 130, p. 265; No. 175, p. 381.

12. D.R., Syiemlieh, 'The Last Days of David Scott', *Proceedings of the North East India History Association*, Fifth Session, Aizawl, 1984, pp. 107-113.

13. D.R. Syiemlieh, 'The Cherrapunji Experiment (1829-1834)', *Proceedings of the North East India History Association,* Fourth Session; Barapani, 1983, pp. 116-123; 'Cherrapunji Versus Darjeeling: The Search for a Sanatorium for Lower Provinces', *Proceedings of the NEIHA*, Sixth Session, Agartala, 1985, pp. 219-225.

14. Dalhousie Muniments GD 45/5/48 No.1.

15. *Ibid.*

16. Robert Lindsay, 'Anecdotes of an Indian Life', in *Lives of the Lindsays or a Memoir of the House of Crawford and Balcarres,* vol. iii, London, 1849, pp. 176-185.

17. Lindsay Papers 9769/30/5/9 volume 2, letter dated 29 October 1786.

18. *Ibid.*, letter dated 3 November 1786.

19. *Ibid.*, 9767/30/5/9 volume 2; letter from Robert Lindsay to Sir John Shore, President of the Board of Revenue, Calcutta, 14 December 1787. The letter is also printed in W. Firminger (ed.), *Sylhet District Records*, vol. ii, Shillong, 1917, No. 301, p. 204.

20. An introduction to this manuscript and its entire text may be read in, D.R. Syiemlieh, 'Leaves from the Past: An Early Account on the Khasis', *Khasi Studies*, vol. v, October-December 1991, No. 3, pp. 86-90.

21. Edward A. Gait, *A History of Assam.*

22. The Memoir has been reprinted, D.R. Syiemlieh (ed.) with an introduction, *Anecdotes of an Indian Life*, North-Eastern Hill University Publications, Shillong, 1997.

23. M.D. Wainwright and Noel Mathews (eds.), *A Guide to Western Manuscripts and Documents in the British Isles Relating to South and South East Asia* , Oxford University Press; London, 1965.

24. A distinction is here made between 'official' government records which are not included in the guide and 'private', covering a variety of records for which the *Guide* referred above provides information.

25. Nightingale Collection, British Museum, Western Manuscipt Section, Addl 30240 B. The manuscript was later published, see E.D. Steel, "On the Khasia Tribe", *Transaction of the Ethnological Society of London*, vol. vii, n.s. 1869, pp. 305-312.The published version has more recently been placed in the very useful portal 'Brahmaputra Studies Database'.

26. For details of this account read D.R. Syiemlieh, 'Leaves from the Past: A Soldier Account on the Khasis', *Khasi Studies*, vol v, July-September 1991, No.2, pp. 36-40.

27. See p. 14 of the original manuscript.

28. John Hugh Morris, *The History of the Welsh Calvinistic Methods' Foreign Mission* , Carnavor, 1910; J. Fortis J Jyrwa, *The Wonderous Works of God*, Shillong, 1980.

29. Nigel Jenkins, 'Thomas Jones and the Lost Book of the Khasis', *The New Welsh Review*, No. 21, 1993, pp. 56-64.

30. Nigel Jenkins, *Gwalia in Khasis:The biggest overseas venture ever sustained by the Welsh,* Gomer Press, Dyfed, 1995. The Indian edition is entitled *Through the Green Doors: Travels Among the Khasis*, Penguin, India, 2001.

31. Andrew May, *Welsh missionaries and British imperialism: The empire of clouds in north-east India,* Manchester University Press, 2012.

32. Alun I.Bannister, "A Past Acquaintance", typescript, Bishopston, February 1993, The author is the grandson of William Pryse, missionary to the Khasis and the people of Cachar-Sylhet, 1850s-1860s.He kindly gave me a copy of the collection of translated letter after attending a lecture I delivered in Edinburgh University in 1994. He was moved to have seen a photo/image of the grave of his grandfather in the illustration of the talk.

33. William Griffith, *Journals of Travels in Assam, Burma, Bootan, Affghanistan and the Neighbouring Countries,* Bishop's College Press, Calcutta; 1847, reprinted 2001, Munshiram Manoharlal Publishers, New Delhi.

34. Jacob Tomlin, *Missionary Journals and Letters*, James Nisbet and Co., London, 1844.

35. Alun I. Bannister, "A Past Acquaintance", Dedication page.

3

Technology and Socio-Economic Linkages of the Khasi-Jaintias in Pre-Colonial Times

Introduction

The history of the Khasi-Jaintias before the arrival of the Ahoms in the Brahmaputra valley is uncertain. Their folklore and oral traditions tell of their supernatural origins, which a section of the tribe have interpreted to explain that they are autochthons of the land now called *Hynniewtrep*. This belief apart, it is generally held that they were one of the first tribal groups to have migrated into their present hills, though when that migration took place and why there was a migration has not been explained. However, what is certain is that this migration preceded that of the Ahoms, as the Ahom chronicles, the Buranjis make no mention of the Khasis in the course of their own migration into the upper Brahmaputra valley, whereas there are references to the Nagas and other peoples the Ahoms interacted with in the course of their settlement in the valley.

When and where the Mongoloid Khasis changed their language to its Austric base is another of those unsolved phase of their past. Their tradition says that they lived for some time in the Brahmaputra valley from where they then settled in the hills that have come to take the name of the tribe. They are said to have first settled in the Jaintia hills moving gradually towards the Khasi hills in the practice of swidden cultivation and search for iron ore. That the tribe was one and whose roots are common is explained in their genealogy and clan structure. The Diengdoh clan of Sohra traces its origin to *Ka* Iaw Iaw, whose mother came from beyond the Kopili. One branch of the clan resided in Jowai where it became known as the Laloo clan. Another branch moved to Nongkhlaw and became the Diengdoh Kylla clan. The fourth branch went to Mawiong and became the Pariong clan.[1] The story of this clan, to take only one example is remarkable for it points to an eastern point from where the tribe came as also it explains the formation of the Khasi clan structure. In all likelihood the tribe was spread over a wide geographical area, from Kamrup where Ahom records[2] mention the names of some of the Khasi *himas,* and south into Sylhet where much of the lowlands toward the river Surma was in the possession of the hill chiefs.[3] Hemmed in by the Garos in the west and the Mikirs in the east whose early histories also need attention, the Khasi-Jaintias were largely confined to their hills where over many centuries of habitation the Khasi ethos took shape.

Periodization

Before we enter into a discussion on Khasi technology, something needs to be said of the periodization of history with particular reference to the hill areas of North East India. Today it is generally accepted to study the past in time divisions, prehistory; ancient; medieval; modern and contemporary. Alternately there have been studies that have not used the time schedule but concepts such as pre-colonial; colonial and post-colonial. Both the patterns referred to are applicable to the Brahmaputra and Barak valleys because all the four divisions may be drawn to interpret the past and particularly because these areas have written material for reconstructing the history from the ancient past. In the case of the hill areas of the North-East such periodizations are not applicable for a number of reasons.[4] While it may be argued that it is possible to reconstruct the pre-history of some of the tribes from material remains, it becomes difficult to provide the tribal histories with 'ancient' and 'medieval' pasts because there is an absence of written material from which to reconstruct that past. Even giving them a modern history becomes difficult because to move into the modern without an intelligent account of the two earlier periods of history would be contributing to a fallacy in history. The difficulties in such a periodization may be overcome as has been suggested by Amalendu Guha. Addressing a seminar at the North Eastern Hill University in 1988, he said:[5]

> In the case of most parts of North East India, periodization breaks down since major segments of our population have lived down to the early 19th century without mastering the art of writing. If we want to know about their pre-18th century past, then the conventional methods of writing history do not help. We need to take resort to research in linguistics, ethnology and archeology for the purpose.

Another problem in attempting an essay on the 'pre-colonial' past of the Khasi-Jaintias is that the time in history covered in the term 'pre-colonial', which, even if it is more apt, covers many centuries and we do not have the material remains or written sources to reconstruct what happened in the post-'pre-historic' stage. Khasi-Jaintia folklore moreover, has not been dated to more judiciously use this source for interpreting the past. It may be possible to use the Assamese and Bengali sources, as the tribe in the hills did not live in splendid isolation, as it is made to believe. There was economic, social and political interaction between the hill people and their neighbours in the plains below. It may be implied, therefore, that over this long period the society under study had first to adjust in the transition from their 'pre-historic' culture to slow developments in time over the next few centuries bringing up their history to the mid-eighteenth century, when material becomes more readily available for studying the tribe and the society.

Sources

This then brings in a discussion on the sources used for this study. There are written materials in the Ahom, Assamese and Bengali languages on the Khasi-Jaintias from which to reconstruct Khasi-Jaintia history.[6] Khasi folklore is also rich in retaining the significant facets of their past In recent times this has become an important tool for historical research on pre-literate societies.[7] However, that these are scattered and cover a long span in. time and referred in sources that have to be verified makes it difficult to arrive at a conclusive understanding of the society. Moreover, there is insufficient material to reconstruct a meaningful history. This is reflected in the existing material on Khasi-Jaintia history where little but very interesting literature has been worked out on their pre-history and pre-colonial history.[8]

This essay will attempt an understanding of Khasi technology and its socio-economic linkages in pre-colonial times just prior to the colonial interaction. It will use material largely in the English language written from around the mid-eighteenth century and till the early nineteenth century. It may be possible to have a broader picture of the Khasi-Jaintia society at this period of time because what the sources indicate is not change, as much as what prevailed at the time the writers made their observations. In taking this course it is not intended to exclude the use of material in the other languages for which references are available. To be candid, the present writer does not have the faculty to use these sources other than through translations or with the assistance of interpreters. Aware of this limitation, the essay will nonetheless largely use material in English, both primary and secondary, together with references in Khasi to substantiate the primary sources used in the presentation.

Agriculture

Agriculture was hardly the mainstay of the early Khasis. As was in vogue among the majority of the hill tribes in the region food was grown by *jhum* cultivation. Discouraging settled agriculture were the high hills with their steep slopes and deep valleys that would have required a great deal of labour for terrace cultivation. Moreover the Khasis did not have use of the plough preferring to till the soil with the *mohkhiew*, the hoe, the shape and name of which Gurdon links with the Burmese and Malays.[9] Jaintias in the plains of what were the Jaintia *Raj* and the valleys in their hills had taken to settled cultivation using the plough.[10] 'The explanation for this technological use over their Khasi neighbours can be found in the Jaintia state structure and exercise of control over much of the Barak valley where the plough was in use, whereas the Khasi *himas* had a control only over the foothills where the plough could not be used. One other factor that might have influenced their agriculture in the hills was that the Khasi *Syiems* controlled the *duars* to the north of their hills opening into Kamrup and the foothills in the south into Sylhet. These lowlands gave them sufficient rice. As one early British report noted: "the Cosseah never cultivates the soil: he always employs Bengalee ryots: he comes down at the time of harvest and carries off the produce."[11] Such was the importance of the foothills that the hillmen went all out to defend their rights over these lands, entering into conflicts for its control with the Ahoms and the Mughal Governor of Sylhet. To further elaborate on this point, extensive trade of produce from the hills, the details of which will follow, bartered for what the plains had in surplus was so established in pre-colonial times, that possibly food in cereal could be more easily procured than grown.

A variety of fruits grew extremely well on the southern slopes of the hills. The Sylhet orange as it was called in Bengal which came from the precipitous slopes of the hills were long famous throughout India while the pineapple was also of superior quality. Pineapple leaves had a utility by the hillmen. These were collected before the onset of the monsoon, soaked in water, pounded and the fibre separated. The fibres were then used for making net pouches or bags, which formed part of the dress of every Khasi. In these pouches they carried their clasp, knives comb, flint steel and betel nut box.

Carriage by Human Labour

The export of fruit, *tezpat* and iron and limestone was all carried by human labour by a variety of specialized baskets. It is to be noted that the Khasi-Jaintias did not have use of the wheel in any form and particularly for the movement of the produce of the hills. Neither is there any reference to the domestication and use of animals. With plentiful forests of bamboo the Khasis became skilled in basketry. The standard basket, the *khoh* was conical in shape, broad and round at the top, narrowing gradually to a point at the bottom. Jaintia baskets usually were upright. Storage baskets with or without lids and measuring baskets were made as also the *kriah* for carriage of tezpat. Other bamboo works were the *prah* for winnowing, the *pdung* a circular and flat tray for cleaning rice, while they used the *knup* and the *trap* as shelter from rain. Mats had a variety of uses, for sleeping on, for drying grain and as a cover to the planks in their houses. Still other specialized bamboo works were the cylindrical frame broad at one end, narrow at the other used for carrying pigs. All weighty and bulky objects were carried on their backs in baskets, with the *star* as a sling distributing the weight from forehead to the back. It must be noted here that all this movement of commodities was performed by human labour. The Khasi are not known to have had use of domesticated animals, nor had they had use of the wheel in any form. Khasi technology was so far behind that one early colonial observer commented: "Man is the only bearer of burdens."[12]

That the Khasis lived in regular villages on fixed sites for generations is an explanation why they have strong attachment to their villages. Villages were not strategically situated, as they did not have fear of each other as some other tribes have of their own kind. Their houses were usually built in the shape of a shell, divided into three compartments consisting of a porch, a kitchen and an inner room. Khasi houses were not large by any standard. This may have been due to the social and family structure which though close-knit lived in individual houses. The material used in the construction was locally available consisting of wood or bamboo and covered with thatch. Walls were made of roughly hewn wood planks for they also did not have use of the saw,[13] though they were experts with the cleaver and the adze. The doors of the houses turned on good wooden pivots. Though iron technology was fairly advanced in the hills, no iron was used in the construction of their houses as they considered it *sang*-taboo. Gurdon makes reference to a number of taboos relating to house construction. Khasis also considered it *sang* to build a house with stonewalls on all four sides; to use more than one kind of timber in building the hearth; to build a house with resinous timber; only the *Syiem* family could use such timber.[14] Why these taboos prevailed is uncertain. What it did have an effect on was the absence of stone and iron in the house construction. Consequently, it has not been possible to study the remains of villages as the material used was perishable and therefore could not stand the test of time. One example of this is the monoliths of Nongkseh village close to present day Upper Shillong. There is no indication of any other human activity in that site. That it was once a village is apparent from the word Nongkseh (*nong*-village-*kseh*-pine tree) and the tradition that the village was situated on the road between the *Bhoi* (northern hills) and the *War* (southern hills).

There are two opinions on whether the Khasi-Jaintias distilled liquor. Khasi tradition says that they did not consume spirits, confining themselves to rice beer. This was what was consumed when the Jaintia *Raja* entertained Robert Lindsay, Collector of Sylhet, when the latter was invited to the hills for a *shikar* and its concluding festivities.[15] Another observer noted that the Khasis distilled spirits,[16] the method of distilling of which was crude to say the least. Fermented rice or millet was boiled in earthen pots and the steam channelled to two outlets to drop into dried gourds. Khasi-Jaintia religious and social practices have made use of distilled liquor, and we are of the opinion that it was a use which was acquired over many years and prior to the more relishing tastes introduced later.

The plastic and graphic arts were conspicuously absent among the tribe. Where there are some drawings on stone should not indicate the general but the exceptional as these are rare findings and will not explain the use of this skill among a large section of the people. The reason perhaps why this was so was the absence of idol worship in the Khasi-Jaintia religion or of erecting totems in their culture, all this despite the use of tools and implements in their iron industry.

Memorial Stones

A striking feature of these hills is the immense number of memorial stones to be found all across the hills. These are not gravestones; they are both irregular and dressed and commemorate the ancestors. It is possible that the Khasis adopted the custom of erecting stones by the force of example or that they started its use when they finally settled in the hills. The art and ceremony was all but lost with only few stones erected in recent times. That the memorial stones are both rough and dressed may indicate that at a point in time there was the absence of the use of iron in extracting and dressing the stones, while the more elaborately done up monoliths give clear signs of the use of the metal on the material. Questions have been raised how these huge stones were transported when the wheel was not used, and how they were erected; they continue to test the minds of archaeologists.[17] While this tradition on stone was indigenous to the Khasi-Jaintias the stone bridges of which there are only few remains have to some extent had influence from Bengal. Thomas Fisher, the Superintendent of Cachar in 1840 mentions the Saracenic style of the bridge at Amwi on the Nartiang-Jaintiapur road which, "quite possible the work may have been constructed by a Mussalman in the employ of the Raja at no very distant past."[18] A unique form of spanning ravines is still to be seen in the southern Khasi Hills. Roots of the rubber and fig trees that have the property to inosculate and form natural grafts have over centuries been used as living bridges.

Iron

The abundance of iron ore in the Khasi Hills made it possible with other requirements of charcoal, clay and water for establishing a flourishing iron industry. It was in part because of the search and location of iron ore that the Khasis moved west from their first settlements in the Jaintia Hills. It is not certain how old this industry could have been, but it is reasonable to say that the Khasis have had knowledge of the excavation and smelting of iron and production of iron implements for centuries. Lt. Yule whose note on the iron industry in the hills was written after extensive travel and observation of the entire process of production had this to say:[19]

> So marked an effect have these works achieved on the undulating hills which cover the country, that in many instances what must once have been like their neighbours, round, swelling knolls, appear to have collapsed and sunk to their skeletons, shewing nothing but fantastic piles of naked boulders; the earth which bound and covered them, having been entirely washed out by the heavy rains following in the track of the miner. So numerous and extensive are the traces of former excavations, that judging by the number at present in progress, one may guess them to have occupied the population for twenty centuries.

In all likelihood the science of excavation, smelting and manufacture of iron came from contact with the plains people with whom there were economic and social connections.

The principal sites for the mining operations were around Mylliem, Nongkrem, Laitlyngkot, around Mairang and Cherrapunji. The open cast mining was invariably worked in the monsoon when with the flow of rain water down the hillsides would assist obtaining the iron ore. After washing the ore a number of times the material would be smelted. Yule then goes on to give a detailed account of the process of smelting the ore. His account differs slightly with that provided by William Cracroft, an early British administrator in the Khasi Hills. His account describes the process using two large upright bellows; the furnace made of pipe-clay braced with iron hoops and the chimney. His account continues:[20]

> At the right side of the bellows and even with the top of the chimney, is a trough containing damp charcoal and iron sand: at every motion of his body the operator with a long spoon tumbles a piece of this charcoal with the iron sand adhering to it, down the funnel of the furnace and when a mass of melted or rather softened iron is formed on the hearth, it is taken out with tongs and beaten with a heavy wooden mallet on a large stone by way of anvil.

The mining and smelting of iron was not a year round operation. Those employed in the industry included women and young boys. The iron ore was often not smelted in the villages adjoining the mines. It was sold in baskets containing three *maunds* of ore, and carried often for many miles to the villages where the smelting furnaces were located. Again, the manufacture of artifacts in the hills was done in workshops differing entirely from the huts in which the first smelting was done. The artifacts manufactured in the hills were swords, arrowheads, spears and the tools for their manufacture.

Robert Lindsay the Superintendent of Sylhet in the 1780s recollected in his memoir that the Khasi Hills produced wood of various kinds, adapted to boat and ship building, and also iron of a very superior quality: "it is brought down from the hills in lumps of adhesive sand, and being put into the forge, produces excellent malleable iron without ever undergoing the process of fusion, the hammer and fire discharging the dross and courser particles at once, thus producing what is called virgin iron, superior to any made in Europe by charcoal.[21]

Much of the iron was sent to markets located in the northern and southern foothills. A reference in Lindsay's memoir explains its transport to the markets in Sylhet:[22]

> I had the gratification of witnessing a caravan arrive from the interior of the mountain, bringing on their shoulders the produce of their hills, consisting of the coarsest silk from the confines of China, fruits of various kinds- but the great staple was iron. In descending the mountain. the tribes descending from rock to rock.In the present instance the only descent was by steps cut in the precipice. The burdens were carried by the women in baskets supported by a belt across their forehead, the men walking by their side, protecting them with their arms.

Iron constituted the principal industry in the hills and the chief export. Limestone too was exported in large quantity though not from as early a date as that of iron. While the hill people made no use of lime, other than as an ingredient in the consumption of *pan*, limestone was exported from the southern hills to the processing villages in Sylhet such as Chunamgunj, to name only one location. It may be assumed that Mughal Bengal first had use of Khasi limestone which was before the East India Company and British traders took interest in this mineral. [23] Limestone was easily exported, as the mines were located close to the numerous rivers flowing into the Sylhet plains. Other commodities for export such as those given in Lindsay's account were carried to the markets by human labour. Imports were few and consisted largely of rice, fish, cotton, silk cloth and salt. These items were traded using barter, for the economy was not a money economy. That the Jaintia and Khyrim *himas* minted coins[24]-*khattra* rupees, need not necessarily imply that there was a circulation of money within these states. Apparently the trade was in favour of the hill people. They appear to have exported more than what was imported. The items of export would have fetched more, even in barter terms to enable the Khasis to convert some of the profits of the trade into gold and silver. This could explain why the Khasi-Jaintias have a fondness for jewellery.

Trade Routes

A trade route from Rahar in Nowgong, through the Jaintia Hills and down to Jaintiapur linked the Jaintia capital in the southern plains to the large village of Nartiang in the hills and the territories in the *duars* adjoining Assam.[25] A less frequented route passed the Khasi Hills. The construction of a road through the Jaintia *hima* was made possible as that state was more advanced in structure and economy. Moreover both end points were within that state's territory. There being many Khasi *himas*, large and small, and without any confederation among them other than at times of war, with poor state resources it was not possible for them to construct a stone paved road going through the Khasi Hills. However it must still have been possible for Khasi traders to have moved commodities from the *duars* opening into Kamrup, going through *hima* Khyrim and the base of Sohpetbneng, up into Nongkseh and Jirang *himas*, passing the larger states of Cherra and Nongkhlaw and down into Sylhet in the description given by Lindsay.

The trade was on established commodities and routes. There do not appear to have been markets in the hills This would be a later development. The markets that were in operation were more in the nature of entreports. The entire trade, again it may be deduced, was in the hands of the hill people just as the workers in the iron industry were. If they so jealously guarded their markets in the foothills, they would not have allowed the trade to get out of their hands. Were the trade in the possession of plains people this would have been noticed. Indeed when Thomas Fisher asked the people he met who they were, the reply he got was that they were "Khyee" - traders.[26]The Khasi-Jaintia *himas* did not have control over the trade other than providing markets which were for the *himas*, more of an expression of their territorial limits.

Through the markets the hill people were able to have political, economic and cultural links with the people of the plains. Some expressions of this interaction are the institutions of *Wahhadadar* among the Shella people, the acceptance of the

Hindu faith and customs among the Jaintia rulers and state officials and among a section of the Khasis in the *War* (the southern slopes of the hills) region; the adoption of a number of Assamese and Bengali loan words in the Khasi language; and their dress, more particularly what the menfolk wore.

Conclusion

The pre-colonial technology of the Jaintias, and we may include the Khasis, in their early state formation, was, to use the words of J.B. Bhattacharjee, "simple, non-industrialised and pre-capitalistic."[27] If in this essay the term industry has been used in making reference to iron, it is to stress the importance of this activity and its organization in Khasi economic life and not attempt to compare it with more organized economic activities both in the region and elsewhere. On all accounts, it appears, the technology used and adapted varied from the crude to the more ingenious.[28] While the Khasis appeared to have adapted some innovations to their economic activity from the technology coming from the plains of Assam and Bengal, they just might have also held back for various reasons; from adapting in increased measure the potential to supply agricultural and forest produce and artifacts to markets in the plains. The technology they applied also appears to have been such that the society was as a consequence stagnant in its economic activity. This stagnation covered a long span in time conveniently understood as their pre-colonial past. The nature of their economy and technology would also provide some clues to why their state structure could not progress beyond their concept and operation of the several Khasi *himas* with their low level of resource mobilization and incapacity to develop into economically and politically stronger states.

Notes and References

*"Technology and Socio-Economic Linkages of the Khasi-Jaintia in Pre-Colonial Times', M. Momin and Cecile Mawlong (eds.), *Society and Economy in North East India,* vol. I, Regency Publications, New Delhi, 2004.pp.22-23.

1. P. R. T. Gurdon, *The Khasis*, reproduced, Low Priced Publications, Delhi, 1996, pp. 63-65.

2. S.K. Bhuyan (ed.), *Jayantia Buranji*, Gauhati, 1937; S.K Bhuyan (ed.), *Deodhai Assam Buranji*,Gauhati, 1933.

3. Suniti Kumar Chatterji, *Kirata Jana Kriti*,The Asiatic Society, Calcutta, 1974, p.166; Syed Murtaza Ali, *History of Jaintia*,Dacca, 1954, pp.I-6

4. Hamlet Bareh has provided a periodization for the history of the Khasis in *The History and Culture of the Khasi People* (revised edition), Shillong, 1985. These are pre-history and early period, medieval and pre-British period and modern history. In like manner B.B. Gupta has divided Naga history into ancient, medieval, modern and cultural histories. See his book, *History of Nagaland*, N. Delhi, 1982.

5. Amalendu Guha, 'Introduction', in J.P. Singh and Gautam Sengupta, (ed.), *Archaeology of North Eastern India*, N. Delhi, 1991, p. 2.

6. The Ahom sources are the several *Buranjis*; in Bengali two of the more important sources are Abdul Aziz, *Jayantia Raiyer Itihas,* Sylhet, 1920 and A.C. Choudhury, *Shrihaiter Itibritta,* Sylhet, 1317 B.S.

7.   P.R.T. Gurdon (op. cit.) provides a number of Khasi folktales which with other published and unpublished stories, social scientists have used to explain Khasi-Jaintia history. See Namita C. Sen Shadap, *The Origin and Early History of the Khasi Synteng People,* Calcutta 1981; Soumen Sen, *Social and State Formation in the Khasi-Jaintia Hills,* N. Delhi, 1985.

8.   Research on Khasi-Jaintia pre-history has been those of Cecile Mawlong, 'Megalithic Monuments in the Khasi-Jaintia Hills: An Ethno-Archaeological Study', unpublished NEHU Ph.D. thesis, 1996, and her article, 'Some Aspects of the Indigenous Earthware of the Khasi-Jaintias Hills', *Proceedings of the North East India History Association* , 19th session, Shillong 1999, pp. 62-68. Marco Mitri, *An Outline of the Neolithic Culture of the Khasi-Jaintia Hills of Meghalaya, India: An Archaeological Investigation* , ISBN, 9781407304632, South Asian Archaeology Series, British Archaeological International Reports, Oxford, 2009.

9.   P.R.T. Gurdon, op. cit., p.12.

10. J.B. Bhattacharjee, 'Brahmanical Myths,Royal Legitimatizing and the Jaintia State Formation', Social and Polity Formation in Pre-Colonial North East India, New Delhi, 1991, p. 97; S.M. Ali, op. cit., p. 80.

11. W.K. Firminger (ed.), *Sylhet District Gazeteer*, vol. III, No. 172. Shillong, 1917 Letter from J.Willes, Collector of Sylhet to Earl Cornwallis, Governor- General, dated Sylhet, 15 September 1789.

12. Thomas C. Watson, 'Chirra Punji and a Detail of some of the Favourable Circumstances which Renders it an Advantageous Site for the Erection of an Iron and Steel Manufactory on an Extensive Scale', *Asiatic Journal*, vol. III, 1834, p. 1.

13. John H. Morris, *The History of the Welsh Calvinistic Methodists' Foreign Mission,* reprinted Delhi, 1996, p. 89, narrates how the Khasis were instructed by Rev. Thomas Jones on the use of a saw:" Hitherto, the Khasis had known no way of securing a plank but by hacking a tree with their hatches; but when the Saheb, by means of his saw obtained several smooth planks from one tree, their admiration was unbounded."

14. Gurdon, op. cit.; p. 159.

15. Robert Lindsay, 'Anecdotes of an Indian Life', *Lives of the Lindsays*, vol. III, London, 1849, p. 178.

16. Alexander Lish, 'A Brief Account of the Khasees,' *The Calcutta Christian Observer*, vol. 7, 1838, p.136. The full text may be read in Brahmaputra Studies Database http://brahmaputra.vjf.cnrs.fr/bdd/spip.php?article98

17. C.R. Clarke, 'The Stone Monuments of the Khasi Hills', *The Journal of the Anthropological Institute of Great Britain and Ireland*, vol.11, 1874, p. 490, wrote that it was highly probable that the method of moving stones on wooden rollers was used in ancient times.

18. Clarke, op. cit., p. 489; Thomas Fisher, 'Memoir of Sylhet, Kachar and the adjoining Districts', *Asiatic Journal*, vol. ix, 1840, p. 834.

19. Lt. Yule, 'Notes on the Iron of the Khasi Hills, for the Museum of Economic Geology', *Journal of the Asiatic Society of Bengal*, No. 129, 1842, p. 854. Also see David R. Syiemlieh, 'Khasi Iron Culture and Iron Trade with Sylhet in the late Eighteenth and Early Nineteenth Centuries', *Proceedings of the North East India History Association,* Eighth Session, Kohima, 1987, pp. 242-250.

20. William Cracroft, 'Smelting of iron in the Kasya Hills', *Journal of the Asiatic Society of Bengal*, vol. 1,1832, pp. 150-151.

21. Robert Lindsay, op. cit., p. 174.

22. So important was this commodity that an agreement was made between the Bengal *Nawab*, Mir Kasim and the East India Company on 10 July 1763, that for a period of five years the *Nawab* represented by his *Fauzdar* and the Company's *Gomastahs* would jointly prepare *chunam*, for which each would defray half the expenses.

23. *Ibid.*, p. 179.

24. Edward A. Gait, 'Some Notes on Jaintia History', *Journal of the Asiatic Society of Bengal*, No.3. 1895, pp. 242-245; Syed Murtaza Ali, op. cit., pp. 88-91, for notes on Jaintia coins and S.K. Bose, 'Symbols in Naranarayan's 'Mudra' and a case of Khyrim coin', *Proceedings of the North East India History Association* , Kohima session, 1987, pp. 92-111.

25. Syed Murtaza Ali, op. cit. Stretches of this road were traversed by J.H. Hutton in 1925. See his article, 'Some Megalithic Works in the Jaintia Hills', *Journal of the Asiatic Society of Bengal,* N.S. vol. xxii, 1926, pp. 333-347. It was along this route that David Scott, Agent to the Governor General North East Frontier, intended to connect Assam and Sylhet by constructing a road in 1826.

26. Thomas Fisher, op. cit., pp. 333-335.

27. J. B. Bhattacharjee, op. cit., p. 101.

28. Cecile Mawlong, "Methods of Preservation Practised in Cherra State: Some Insights," *Proceedings of the North East India History Association,* Agartala Session, 1997, pp. 101-106.

4

Control of the Foothills: Khasi-Jaintia Trade and Markets in the Late Eighteenth Century

This essay will take up a discussion of the two issues of trade and markets in the foothills of the Khasi-Jaintia in the late eighteenth century. In the discussion the focus will not be so much on the economic dimension of that past and activity but the political implications of the control by the Khasi-Jaintias of the foothills and the trade and markets located therein. By this we may then arrive at an understanding of the significance of the border markets in Khasi-Jaintia life and the impact these markets had on the community at large. When this is established, it will then be possible to take up in another essay, an account of the markets and border trade under colonial rule.

Territorial Limits

It is uncertain when and why the Khasis left the Assam plains after their migration into what is today North East India. In the process of taking possession of the hills the Khasis evolved their own social and state structure. They first moved in the southern direction into what was to become the Sutnga *hima;* there from they moved both south into the plains of the Surma valley and west into the central highlands of old Shillong and Sohra *himas and* further towards the westernmost part of their settlement in the *himas* of Nongkhlaw, Rambrai, Nongspung, Jirang and Nongstoin. Though there is no sequence known of this migration and settlement that the settlement was from east to west can be substantiated by the advance of their technology in erecting of monoliths and the dressing of some of the larger stone remains and the oral tradition that the move westward was in part triggered by the search for iron.[1]

While the state formation process was relatively far advanced' in the Jaintia *hima*, state formation in the several Khasi *himas* had not progressed evenly. Tradition tells of the existence of thirty Khasi *himas* and twelve *Dollois* in pre-colonial times, It is understood that the term *Khadar Dolloi* is with reference to the twelve *Dollois* of the Jaintia State and the number of these functionaries who assisted in the administration of that *hima*. The term of the official has an Assamese origin suggests that the Jaintia *Rajas* borrowed the term and perhaps their functions and applied it to the administration of their *hima*. That there were numerous Khasi *himas* prior to the eighteenth century there is certainty. But whether there were actually thirty *Syiemships* is a matter of contention.[2] The state formation process of these *himas* was somewhat different to Jaintia, in that many were very small to small *himas;* not all were directly in contact with either of the peoples of the Brahmaputra and Surma valleys and not all had Brahmanical influences as Jaintia and some of the *himas* in the *War,* the southern region were to experience. Several of the *himas* were large enough to have included in their territories the foothills to the north and south. Eight Khasi chiefs of whom we may note Shillong, Jirang, Myriaw and Rambrai *himas* had control of the *duars* entering from the hills into Kamrup. Across the hills the larger states that had land in the Sylhet plains were Sohra, Maharam and Nongstoin. Shella, a much smaller *hima* comprising a number of villages was located on the very foothills.

An element nurturing the more developed state formation process in Jaintia, the easternmost of the "Khasi" states was the Brahmanical myth of the origin of the *hima* which J.B. Bhattacharjee has studied in detail.[3] This apart it was also the only hill state to have had control of the Sylhet plains beyond the foothills, extending as far in to the plains up to the river Surma. Another factor for the more advanced nature of the Jaintia state was the size of this *hima*. It abutted on the Karbi hills further east, the foothills in Nowgong and Kamrup, running across the entire hill region of the Meghalaya plateau and well into the Sylhet plains. Of the control of the foothills to the north and south of the Jaintia State, more will be said presently.

Border *Hats*

Markets were largely located in the foothills. The importance of the markets in Khasi-Jaintia economy is illustrated in the tribe naming the days of the week after the weekly markets. The Khasi markets in the hills would be a later development. The large and more important *hats* in the southern foothills were Pandua, of which Robert Lindsay, one of the early Superintendents of Sylhet gives a detailed description,[4] Punatit in the Laur Pargana and Jaintiapur. Of these Jaintiapur was a flourishing mart, located well into the plains and closer to the river Surma.[5] It was linked with a road that connected the Jaintia capital with the hill territory and went on towards Rahar in Nowgong. Pandua located on a hillock close to Shella, was a mart "where the Bengal, Assam and Garrow goods are bought and sold."[6] On the Assam foothills of the Khasi-Jaintia *himas* were located the Nauduar, the nine passes opening the hills to Kamrup. Some of the principal markets located here were Barduar, Rani (under the management of a vassal of the Khasi *Syiem* of Nongkhlaw) and Sonapur.

We have earlier said that there did not appear to have been markets in the hills.[7] The political and economic forces which would change this situation and enable markets to open within the hills (to be discussed in the third part of the series on markets and trade) had not developed even in the later part of the eighteenth century. Throughout the later pre-colonial times, the Khasi-Jaintias controlled the foothills to such an extent that these were to become the territorial limits of the several *himas* towards the northern and southern plains. The absence of markets in the hills however did not prevent cross-hill trade. Their control over the several *duars* and the markets located in these openings into the northern plains and the numerous *hats* located likewise in the southern foothills enabled Khasi traders to shift commodities from one valley to another. The markets in the foothills mentioned were of two types: They were held periodically to enable the Khasi traders to exchange their produce. Some others were more in the nature of entreports if the commodities such as *muga,* fruits and iron were to be sold in markets across the foothills. Pre-colonial Khasi-Jaintia economy was not a money economy. The Jaintia Rajas were known to have minted and circulated to a limited extent *Khatta* rupees. Cowries were in circulation in Sylhet and some exchange could have been transacted using this form of money but to a limited extent. The commodities exported from the hills, of which iron was the single largest export and the import principally of rice into the hills was made possible by the surplus exchange of exports from the hills through barter. Whatever material was moved through the hills from *hats* in either of the foothills appear to have been in the control of the Khasis. There is no reference to the people of the Assam and Bengal plains being involved in the cross-plateau trade. However, it cannot be ruled out that they were not involved in the exchange of commodities.

The controls of the lands in the northern foothills by eight Khasi chiefs' were in the nature as fiefs from the Ahom monarch. They paid nominal allegiance for the control of these passes. The policy that the Ahom rulers followed towards these hill chiefs was one of conciliation. With a view to disarm the frequent inroads of the hillmen into the Assam plains, they admitted them to a share in the produce of the soil, a policy which was also pursued towards the tribes bordering the northern hills of the Assam valley. The explanation of this policy is explained in part in the geographical position of Assam and the extreme difficulty the rulers faced in defending a valley of four-hundred miles length from the inroads of the hillmen.[8]

Assertion for Control of the Southern Foothills

That Ahom-Khasi relations were quite cordial is apparent as there are very few references of the more dominant Ahoms attempting to extend their rule over these hill tribes. The Khasi-Jaintias however appeared to have more political, social and economic interaction with their neighbours in the southern plains. The Bengal *Nawabs* like their Ahom counterparts did not disturb the hill people from exercising control over the foothill hats and the business conducted therein.

The trade in iron and limestone was well under way by the time the East India Company was granted the Diwani of Bengal in 1765. The colonial authorities immediately applied a boundary, a hill-plain divide where previously there was none. Problems then very naturally arose over the control of the foothills. The Company believed their right extended to the foothills and the Sylhet plains which was a natural extension of the Bengal plains. New revenue and administrative arrangements for Bengal quickly affected the Jaintia Raja's position. The East India Company officials, many of whom were engaged in private trade, questioned his collection of tolls on all boats plying on the upper reaches of the river Surma. They believed that their private trade too should have the benefits of toll exemption as the Indian traders were enjoying. It was over this issue of toll and control of the Surma that the Calcutta Council decided to exert military pressure over the hill chief. Proceeding from Sylhet on 24 March 1774 with three companies of *sepoys*, Captain Ellerker was ordered to move towards Jaintiapur reaching the place after five days' march. There was one small skirmish enroute and another sharp engagement at Jaintiapur before Raja Chatra Singh went into the hills. Peace was eventually made when the Raja signed a treaty with the Company, on 12 June 1774, promising to pay the Company Rs. 15000 as compensation for the cost of the expedition and that there should be free and unimpeded navigation on the river Surma.[9]

Following the short engagement in Jaintiapur came a more serious and protracted problem for the Khasis. As they were now entering into both trade and political relations with the Company officials, they were not free to exercise their territorial claims over the foothills. Consequently after 1765 the Khasis are reported to have made several raids into the Sylhet plains. The Khasi reason why they made raids is unexplained. However, we may deduce that these raids came as a response of the hillmen finding their *hats* and their control over the lowland in Sylhet passing to the Company administration. Experience showed Robert Lindsay, the Sylhet District Collector, whenever the Khasis raided the plains, to immediately follow the Khasi confederate chiefs into the hills to keep them out of bounds. "This policy", he wrote, "was never attempted during the Mogul Government, but I found it attended with every good affect I wished for."[10] To prevent further raids several Khasi chiefs were given rent free lands in the plains of Sylhet with the intention that this would act as deterrents against raids by the hill people. Holland, Lindsay's predecessor as District Collector of Sylhet granted such lands called *tunkhwahs* (land granted to a chief for military service) among others to Oboo Singh of Mawsmai and Soubu Singh of Cherrapunji. Lindsay later appointed a "Bengali-Khasi", Baroo, Choudhury over a small pargana of Shamnagar and put under him several *ghats* between Solegur and Chattack to protect that part of the river Surma.[11]

The first serious attack on Sylhet occurred in 1783 when the hillmen demanded the head of a *havildar* who had treated them with contempt. Pandua was attacked with considerable loss to both sides. The *havildar* who had got the wrath of the hillmen was enforcing an order of the local Company authorities prohibiting the collection of the toll from the markets on the foothills. That the Company authorities recognised the collection of tolls by the Khasi chiefs is evident in the advise an officer gave to his Dacca Council that markets should not be set up "close to the hills where Cosseahs collect musool."[12] Four years later in 1787, the Khasis made another bid to exert control over the foothills-and their parganas in Sylhet. Ganga Singh, the *Syiem* of Shella, led the reprisal against the Company.[13] Following the attack it was suggested by an official of the Company that the boundary be, "closely defined within which we should not admit the Hill Rajahs to exercise the smallest authority, though we might cultivate their friendship, and give every encouragement to their people to come amongst us unarmed, and settle as subjects and ryotts."[14] Willes, the Collector, wrote to Lord Cornwallis, the Governor-General, in more detail."[15]

> I think the policy to be pursued with all these hill people is first to establish the several limits, and then whenever the occasion comes to support our own authority within our own territories, and not, as at present, having lands nominally the Company's partly subject to us and partly subject to the Cosseah, in which they exert authority.

The hill chiefs again attacked Pandua in June 1789. The provocation was the confinement by the Sylhet administration of a Bengali supporter of the Khasi chiefs who held *tunkhwah* lands in the plains. Soon very many villages -137 it is reported. joined in the challenge to increased British control over the foothills.[16]

The Khasi raids into the plains necessitated for the Company more firm control over the foothills and the framing of a policy to direct the Company's interests in Sylhet with the Khasis. On the recommendation of Willes, the Governor General-in-Council instructed that the hill Khasis should not be permitted to hold any land in the plains within the Company's limits, either as proprietors or farmers or under and tenure whatsoever. The Khasis were to be allowed free movement in the district for the purpose of trade provided they descended unarmed and conducted themselves peaceably. Regulation 1 of 1790 was subsequently incorporated into the Company statute book. This regulated the Company trade in Sylhet and restricting the movement of British subjects beyond the north west of the Surma.[17] As a further step to prevent the hill people raiding the plains, it was proposed in 1799 to survey the foothills and to demarcate the border with the Khasi hills along which a line of forts were established.[18]

Conclusion

In all this narration the perspective has largely been drawn from literature which provide the British view of things. Were the Khasis able to have expressed their grievances in written form apart, from their use of force, or had the colonial power the opportunity to know the Khasi view of things, the reasons for their disturbed situation would have been better understood. We can only deduce that the Khasis were disturbed that the expanding British power was affecting their long established rights over the border *hats* and the collection of tolls. One other factor must also have been of concern for the Khasis. The Sylhet plains supplied much of their requirement of rice. An effective method used by the British in retaliating the raids into the Sylhet plains was to prevent rice and salt from going up into the hills. This same tactic would be later be followed by the British in their taking control over the Jaintias, Garos and other hill tribes of the region. The colonial power had early realised that if the markets on the foothills were closed and their entry restricted into the foothills, the hill chiefs would not only have lost their political control over much of the plains, it would also have closed their markets and their source of revenue thereby making them more dependent on the colonial power.

Notes and References

*'Control of the Foothills: Khasi-Jaintia Trade and Markets in the Late Eighteenth Century,' F. A Qadri (ed.), *Society and Economy in North-East India*, vol. 2, Regency Publications, New Delhi, 2006,pp. 326-334.

1. For details of this early history of state formation see the article of B. Pakem, "State Formation in Pre-Colonial Jaintia, Surajit Sinha (ed.), *Tribal Polities and State Formation in Pre-Colonial Eastern and North Eastern India*, Calcutta, 1987 pp.243-249; also read J.B. Bhattacharjee, 'Brahmanical Myths, Royal Legitimation and the Jaintia State Formation,' chapter 5, *Social and Polity Formations in Pre-Colonial North East India: The Barak Valley Experience*, Har-Anand Publishers, New Delhi, 1991; Homiwell Lyngdoh, *Ki Syiem Khasi Bad Synteng*, Shillong, 1964, and Hamlet Bareh, *The History and Culture of the Khasi People*, Shillong, 1985.

2. Refer to the memoranda of Hamlet Bareh and R.S. Lyngdoh in R.T. Rymbai, *Land Reform Commission for Khasi Hills*, Shillong, 1974, p. 58, p. 97.

3. J.B.Bhattacharjee, 'Brahmanical Myths, Royal Legitimation and the Jaintia State Formation', *Social and Polity Formation in Pre-Colonial North East India*, New Delhi, 1991,p.97; Syed Murtaza Ali, *History of Jaintia*, Dacca, 1954, pp.1-6.

4. Robert Lindsay, *Anecdotes of an Indian Life*, edited with an introduction by D.R. Syiemlieh, NEHU Publications, Shillong, 1999.

5. Syed Murtaza Ali, *History of Jaintia*, Dacca, 1954. Till date this book provides the most detailed history of Jaintiapur.

6. R. H. Phillimore (comp.), *Historical Records of the Survey of India: 18th Century*, vol.1, Dehra Dun,1945, p.115.

7. D.R. Syiemlieh, 'Technology and Socio-Economic Linkages of the Khasi-Jaintias in Pre-Colonial times,' M Momin and Cecile Mawlong, (eds.), *Society and Economy in North-East India*, Vol. 1, Regency Publications, New Delhi, p. 31.

8. National Archives of India, Foreign Political Consultations, 21 December 1835, No. 16; West Bengal Archives, Bengal Political Consultations, 24 November 1835, No. 17.

9. For details see, D.R. Syiemlieh, *British Administration in Meghalaya: Policy and Pattern*, Heritage Publishers, New Delhi, 1989, pp. 16-18.

10. Walter K. Firminger; *Sylhet District Records*, vol. iii, Shillong, 1917, No.3.

11. *Ibid.*, No. 119.

12. *Ibid.*, vol.iii, No. 172.

13. For more details of this and other raids by Khasis into the foothills read, D.R. Syiemlieh, *British Administration in Meghalaya: Policy and Pattern*, Heritage Publishers, New Delhi, 1989, pp.16-24 and P. N. Dutta, *Impact of the West on the Khasis and Jaintias:A Survey of the Political Economic and Social Change*, New Delhi, 1982, pp.28-42; Gunnel Cederlof, *Founding an Empire on India's North-Eastern Frontiers: 1790-1840: Climate, Commerce, Polity*, Oxford University Press, New Delhi, 2014, pp.49-52

14. Walter K. Firminger, *Sylhet District Records*, vol.iii, No. 119.

15. *Ibid.*, vol.iii, No. 119.

16. *Ibid.*, Nos.146, 172; B.C. Allen, *Assam District Gazetteer: Sylhet*, vol. ii, Calcutta, 1905, p. 35.

17. W.S. Seton-Karr, *Selections from the Calcutta Gazette*, vol. ii, pp. 30-31.

18. B.C. Allen, op. cit., p. 36-37; P. R. T. Gurdon, *The Khasis*, 'Introduction,' reproduced, Low Price Publications, New Delhi, 1996, p. xvii; Helen Giri, *The Khasis Under British Rule 1824-1947*, Regency Publications, New Delhi, pp. 37-38.

5

Trade and Markets in the Khasi Jaintia Hills: Changed Conditions in the 19th and 20th Centuries

"The development of the shop can be traced in the Khasi and Jaintia Hills from its very earliest beginnings".

Introduction

It is fairly established that the Khasi-Jaintias were a society who had engaged themselves in trade and commerce over centuries. They have the advantages of living in a land endowed with mineral resources which could be exploited; with salubrious climate; varied vegetation and located in a geographical area which came to their gain in trade. Trade was largely with the Brahmaputra and Surma valleys. It covered a wide range of exports produced in the hills including fruits and spices, pan, areca nut, limestone and iron. The import of commodities the hill people did not produce was largely salt, rice and cloth. Certain items of trade such as *muga*-silk originated in the northern plains and was taken across the hill section of the plateau and sold in the southern plains. We do not have reference to any organized markets in the hills in the pre-colonial period though there could have been some exchange of goods from one village to another or across the Bhoi, Khynriam and *War* areas of the land.[1] An interesting feature of the village formation and settlement in the hills was that villages were located largely in the Khynriam region and the southern foothills. It may be assumed that apart from other reasons, the proximity of the villages to the trade centres to the south was a motivating factor for the location of the villages. The two known trade routes from Assam and Sylhet through the hills passed by some of the large settlements such as Nartiang, Jowai and Jaintiapur on the Jaintia side and Nongkseh and Sohra on the route through the Khasi hills.

The cross plateau and frontier trade was largely in the hands of the hill people who jealously asserted their claims and control of the trade and markets. There are numerous references in Bengali and Ahom literature of the markets and trade conducted in the foothills of then Khasi-Jaintia Hills. There are several descriptions of the Khasi markets and trade in early British accounts. Alexander Lish, a missionary of the English Baptist Mission of Serampore who resided in Cherrapunji (now Sohra), from 1832 to1836, wrote a long account of the Khasis, in which he enquired how the people lived. He wrote:[2]

> The Khasis have always been in the habit of bartering the spontaneous productions of the hills for those of the plains. Oranges, honey, iron, bee's wax, ivory, Indian rubber, these they give in exchange for rice, fish, salt, but more frequently for specie. Fruits and grains of different kinds, with potatoes grown in the interior and in the valleys are brought by the inhabitants to the principal markets in the hills and are also taken to the plains.

He continues his narrative:[3]

> Considerable intercourse is likewise carried on by the Khasees with the Assamese, by whom they are supplied with cloths of different kinds, such as the moonga commonly worn by them and various coloured and flowered silks which are highly prized by the Khasees. Limestone which abounds in the hills is another course of profit for the Khasees. Lime is burnt to a considerable extent on the banks of the Soormah and brought down to Calcutta and Dacca. But their greatest profit has, till of late years, been derived from their iron works. The digging, washing and smelting of the ore, employ many besides the gains it brings to the masters of the works. They manufacture their own swords, hatchets axes and fit their own arrows.

Lish makes a note of the Khasis exchanging their produce for specie which in all probability was gold if the jewellery of the Khasi women was any indication. However, the volume of trade is not assessed as there are few references to this aspect of the trade. That the limestone trade had been in operation for perhaps a century is noted.[4] The iron trade from all accounts seems to suggest that this industry preceded limestone extraction and processing.[5] The variety of traded commodities would indicate the trade was not insignificant.[6]

Another reference of a market in the hills is to be found in the journal of William Griffiths. Griffiths was Superintendent of the East India Company's Botanic Gardens at Calcutta. He noted an entry in his diary of 1835 while on a botanic collection in the Khasi and Jaintia hills:[7]

November 12th.-Nurteng (Nartiang) occupies a large place in these hills, perhaps next to Joowye(Jowai),.The market, which took place today, is outside the village and close to our bungalow: it is well attended, but the amount of persons could not exceed 100 to 200, and those form a considerable amount of all the persons capable of bearing burdens from the neighbouring villages. The luxuries exhibited are all Khasyan (Khasi), consisting of stinking fish, some other things of dubious appearance and still more dubious odour, millet and the inferior grains, and the fashionable articles of Khasya clothing and the adjuncts to that abominable habit pawm chewing. There was plenty of noise, but still order prevailed: no other rupees than the *rajah's* were taken, and even pice were refused. Iron implements of husbandry of native manufacture were vended; in short all the various luxuries or necessities of a Khasya are obtainable.

Trade and Money Economy

The pattern of trade outlined in the references above, continued till the third decade of the 19th century. By then British colonial interest in political control and with it some trade interest and economic control was becoming evident. British control of Sylhet after 1765 and Assam after 1826 meant that the Khasis has perforce to negotiate with the new political masters. The Khasi-Jaintia *Syiems* soon realized that their control over the foothills and border markets had undergone change and though they were still in control of the markets located there; these no longer signified their independent control over the *hats* and the hinterland. Though border *hats* continued to trade by barter for sometime more, they had lost their position as indications of the extent of the geographical limits of the Khasi *himas*. Boundaries were demarcated first in 1790 and 1799 and plotted in maps beginning with Fisher's map of the independent Khasi estates 1829.[8] As British imperialism expanded and took firm control after the Khasi resistance to British rule 1829-1833, and the annexation of the Jaintia hills and *parganas* in 1835, boundaries were further defined. The new relationship affected in large part the pattern of trade, its volume and the nature of the markets.

Money economy was introduced but gradually in the hills after the British assumed political control over the Khasi and Jaintia Hills. Barter trade however continued well into the mid 19th century, particularly in the exchange of those commodities which the European and Bengali traders were not engaged in – fruits, spices, iron to name a few. The use of money and money economy came in slowly in the hills. It had started with the use of cowries as a medium of exchange and tax in Sylhet in the early part of British rule in that district. By 1835 the Company's silver rupee had become the legal tender in British India.[9] The Company legal tender was also used in the "natives states" with whom the colonial state had political relations.

Limestone and Iron

British colonial rule brought about significant change in the economy of the Khasi-Jaintias. The introduction of new vegetables and fruits by the British added a new dimension to the economy.[10] Hitherto agriculture had little relevance to the tribe. The cultivation of potatoes for instance first introduced in the hills by David Scott in 1831 dramatically changed the purpose of production largely for markets beyond the hills.

The annual production was estimated in 1853 at 30,000 *maunds*.[11]Joseph Hooker the botanist, mentioned that within twenty years of its introduction, Calcutta was supplied potatoes from the Khasi hills.[12] W. J. Allen who made a report on the administration of the Khasi and Jaintia hills in 1858 mentioned an attempt by a Khasi trader to market potatoes in Calcutta.[13] The cultivation of the tuber was initially confined the central portion of the hills. The demand had grown so enormously in Calcutta that Allen noticed an increase in area in the cultivation of the crop in the Jaintia Hills and in the west towards Maharam *hima*. 50,00 *maunds* were reported to have been exported in 1858.[14] By 1876-1877 the production had increased to enable an export of 200,500 *maunds*.[15] Similarly the demand in Calcutta and Bengal for the hill oranges resulted in increased cultivation of the fruit. The trade in oranges was for some time almost entirely in the hands of Henry Inglis, the Assistant to the Political Agent of the Cherra Political Agency. He was able to secure leases on almost all the orange groves in the southern hills. Pineapple, *pan, tezpat* arecanut and some amount of cotton grown on the northern foothills continued to be exported.[16] Though varieties of rice were grown in the hills,[17] the grain was the largest of the imports from Sylhet.[18] As is evident from the account so far, the greater portion of the trade in the Khasi-Jaintia hills was carried on with Sylhet. Trade with Assam was comparatively insignificant being largely a barter trade.[19] Trade with Sylhet was about equally balanced between what went out and the imports into the hills.[20]

British intervention in the economic activity of these hill brought them substantial revenue and profits for individual traders. The colonial government first under the East India Company and later under the British Indian state derived revenue from the limestone trade. Little revenue appeared to have been collected before 1853, when the amount principally from rents of limestone quarries was Rs 1047. By 1858 revenue collection increased to Rs 23,023 as a result of more extensive working of the limestone quarries. By 1877-1878 rent from limestone quarries fetched the government revenue of Rs 66,963.[21] Henry Inglis' Company had a monopoly of the limestone trade through the 1840s and into the 1880s. Armenians and several Indian lessee holders were also involved in this trade.[22] As in the previous century the trade largely supplied Bengal with its requirement of lime. Coal was another mineral the British secured rights over. Coal was first discovered in and around Sohra in 1832. However, the vast deposits of coal could not be worked profitably as the expense of carriage prevented it being of much commercial importance. Coal was extracted and sold commercially from its locations in Sohra and Lakadong in the Jaintia Hills.[23]

If the working and capital investment in limestone and coal was largely European, the iron industry was entirely in the hands of the Khasis. We have made reference to this industry in an earlier volume of this series.[24] Here we may go into the details of the trade and marketing of iron in the period under review. The principal sites for the mining operations were Mylliem, Nongkrem, Laitlyngkot, Nogundee and the region around Sohra. The open mines were worked during the monsoon to take advantage of the rains to unloosen the iron nodules. Water was also required for its collection in troughs. From these mining sites the iron was carried to the smelting furnaces which might not have been far. There are several descriptions of this smelting process and sketches of the furnaces.[25] The iron industry must have been large and spread over a number of villages. Joseph Hooker was moved to write that from the summit of the Kyllang Rock in the west Khasi hills " the tingling sound of hammers from the distant forges on all sides was singularly musical and pleasing; they fell on the ear like "bells upon the wind," each ring being exquisitely melodious, and chiming harmoniously with the others."[26]

The pig iron in lumps were then carried to the markets in the southern foothills not very different from what Robert Linday, the Superintendent of Sylhet had described in the late 18th century.[27] At Pandua iron was sold in lumps called ' biri' at Re 1 and 5 *annas* per *maund*.[28]In Nongkrem iron was sold at Re 1 and 2 *annas* a score, about a dozen pieces went to the *maund*. At Pandua they were sold by weight at Re 1 and 4 *annas* a *maund*.[29] At Chattack iron could be purchased at Rs 1 and 4 *annas* to Re 1 and 6 *annas* per *maund*. Better iron which was beaten into bars called 'peti' was sold for Re 1 and 10 *annas* to Re 1 and 12 *annas* per *maund*.[30] Over a period of fifty years 1829-1879, the cost of iron in Sylhet was relatively stable. The cost of carriage from the hills to the markets in the plains was about 6 *annas* per *maund*. The Khasi traders who had a control over this trade got on average profit of only 2 *annas* per *maund* sold.[31] It is estimated that in 1858 between 45,000 to 50,000 *maunds* of iron was exported valued at Rs. 67,000 and more.[32] In 1876 -1877 Rs 7000 worth of iron implements were exported from the hills against nails and ironmongery imported to the value of Rs 18,000.[33]

Bengali blacksmiths of Sylhet preferred Khasi to other iron produced in the region because of its malleability. The iron brought down to Sylhet underwent a second fusion to remove impurities after which it was fashioned into agricultural implements and put to a number of uses including double hook-like nails for fastening planks for the ship and boat built in the district. Interestingly of the boats constructed, a variety called *barki* were almost exclusively used to transport limestone from the base of the Khasi hills to the processing depots in the plains. The import of cheaper English iron into India and the region was the cause of the near collapse by the 1870s of the Khasi iron production and trade.[34] Remnants of this industry can be seen even today in and around Mylliem. The blacksmiths continue to fashion implements and the like using technology and designs that go back several centuries.

Markets

The principal markets at the foot of the hills on the Sylhet side were Bholagunj, Chattack, Lakhat, Jaintiapur, Jaflong, Pharalbazar, Maodong, Ponatit, Molagul and Lengjut.The markets were held at regular intervals of eight days to enable the hillmen to visit the different *hats* in rotation. The *Syiems* of Sohra and Khyrim levied market dues at Lakhat and Bholagunj respectively.[35]Other *Syiems* with land towards Sylhet also had control over the markets from which they derived *khrong,* a levy. The trade in these border *hats* was largely in the hands of the Khasi- Jaintias and a few Bengali traders. The latter called "box *wallahs*"[36] were amongst the first on the traders from the plains to set up shop in the hills after British colonial state had exerted its political control over the several Khasi *himas* and the Jaintia *Raj*. In all probability these were the traders, John C. Thornton, a medical doctor makes mentions of grain dealers, oil sellers and other petty trades' people who supplied the wants of the soldiers stationed in Sohra in the early1860s.[37] The entrepot markets towards the northern foothills were located in Gobha, Rahar, Sonapur, Rani, Bardwar and Boko among others located at the *duars* opening into Kamrup and Nowgong districts of Assam.[38] Bengali and later Marwari were amongst the first of traders to set up shop in Shillong soon after its foundation in 1866. The growth of urban centres at Sohra, Shillong and Jowai and the requirement of the provincial and district administrations to staff the administrative and military and police services further encouraged the economic activity of the Khasi, Jaintia, Bengali and Marwari traders to meet the requirements of the town people and supplying other material to the smaller *hats*.

Such a prospering economy could not have been possible without a more extensive network of road communication. There were eight principal roads in the district in the 1870s. The main road connected Shillong with Gauhati which was opened to wheeled traffic throughout is length in 1877. The road built by David Scott in 1828-1831 traversing the entire hills from south to the north continued to be in use to the Khasi *himas* of Sohra, Mawphlang, Langrin and Nongkhlaw which were required to meet part of the expenditure for its repair. Other roads led to Mawphlang, to Nongstoin towards the Garo hills, to Jowai and on to Jaintiapur in Sylhet and another from the subdivision headquarters to Nowgong. A good road connected Shillong to Sohra.[39]

From small shops that were the beginnings, there grew a number of markets in the hills. B C. Allen makes an interesting observation in the official gazetteer: [40]

> The development of the shop can be traced in the Khasi and Jaintia Hills from its very earliest beginnings. Along certain roads there is always a large traffic on market days, and an enterprising women takes her seat with a basket full of goods at the roadside. ... If her undertaking proves remunerative she builds a little shelter. Yet none the less they form one end of the scale of trade whose higher notes are represented by Liberty or Harrod's stores.

By the turn of the 19th century there were eight principal markets in the Khasi hills and eight markets in the Jaintia Hills with markets held at different dates in the *War*, the southern foothills, where it is by now apparent much of the trade and commerce was located. The Khasis had an eight day week because the markets were usually held every eighth day. The location of the markets gave the names of the days of the week. The Khasi markets were located at Lynkat at Barapani or Khawang ; Nongkrem ; Mawlong, a mart at Laban, Shillong; Ranghep held at different dates in one of three locations at Sohra, Mawtawar close to Shillong and Nongkhlaw; Shillong and simultaneously at Laitlyngkot ; Pomtiah, a small market in Mawkhar in Shillong; Umnih; and Iiewduh the largest market in the hills located in Shillong.[41] The Jaintias held and named their markets after *hat* Jaintiapur; Khyllaw at Sutnga; Pynsing ; Mawlong at Nartiang ; Musiang at Jowai; Muchai at Shangpung; Pynkhatat Mynsoo and Thymblien.[42]

Conclusion

It is interesting to note that though the political and administrative connections of the hills were increasingly connected with Assam, the economic links were initially stronger with the Surma valley. The construction of the Shillong -Gauhati road and beginnings of motor vehicles for passenger and goods in the early 20th century to some extent increased the economic activity on this route.[43] The partition of the subcontinent and the tumultuous developments with the collapse of colonial rule further increased the dependence on the northern route for trade and commerce. Fortunately the facilities for communication were in place by 1947. Partition on the other hand severely affected the trade and *hats* in the southern hills. Overnight a new boundary and an international one at that was erected which almost closed the once flourishing trade. The fallout of this had its affects on the trade and commerce of the southern sector of the hills. Over the past half century attention has increasingly been drawn to the markets higher up and towards Shillong and Jowai. The dislocation of the business that had been in operation for generation has so affected the *War* people that large numbers have relocated their *pan* and arecanut business and started other trades in the *Bhoi* area. What were once border markets in the days of Khasi- Jaintia control over the marts continue as border *hats* today. The changed situation however has not been of advantage to the people.

Notes and References

*David R Syiemlieh, 'Trade and Markets in the Khasi Jaintia Hills: Changed Conditions in the 19th and 20th Centuries,' D. R. Syiemlieh and Manorama Sharma (eds.) *Society and Economy in North East India*,vol.3, Regency Publications, New Delhi, 2008.pp.51-61.

1. *Bhoi*-the northern; *Khynriam*- the upland; *War*- the southern parts of the Khasi- Jaintia hills.

2. Alexander Lish, 'A Brief account of the Khasees,' *Calcutta Christian Observer,* 1838, p.138.

3. *Ibid*.

4. See for instance the numerous references in Robert Lindsay, *Anecdotes of an Indian Life*, with an introduction by David R. Syiemlieh, North Eastern Hill University Publications, Shillong,1997.

5. David R. Syiemlieh, *British Administration in Meghalaya: Policy and Pattern*, Heritage Publishers, New Delhi, 1989, pp.8-9.

6. David R. Syiemlieh, op. cit., 98-102

7. William Griffiths, *Journal of Travels in Assam, Burma, Bootan, Affghanistan and their Neighbouring Countries*, Bishop's Cotton Press, Calcutta, 1847, p. 169.

8. For a general description of the survey of Sylhet and the hills see R. H. Phillimore (comp.), *Historical Records of the Survey of India, 1815-1830*, vol. II, Survey of India, Dehra Dun, pp.49-52.

9. Dharma Kumar (ed.), *The Cambridge Economic History of India*, vol. II c. 1757-c. 1970, Orient Longmans in association with Cambridge University Press, Delhi, 1984, pp. 768-769.

10. David Scott, Agent to the Governor- General North East Frontier wrote to his friend Thomas Watson, that he was taking up with him to Cherrapunji some plum and apricot trees. He makes mention of planting potatoes, turnips and beet roots. David R. Syiemlieh, *British Administration in Meghalaya: Policy and Pattern*, Heritage Publishers, New Delhi,1989, p. 52.

11. A. J. M. Mills, *Report on the Khasi and Jaintia Hills 1853*, reprinted, North-Eastern Hill University Publication, Shillong, pp. 3, 37-38.

12. Joseph D. Hooker, *Himalayan Journal*, vol. ii, Today and Tomorrow Printers and Publishers, New Delhi, p. 278.

13. W. J. Allen, *Report on the Cossyah and Jynteah Hill Territory*, Calcutta, 1858, reprinted Shillong 1903, p.50.

14. *Ibid*.p.49.

15. W.W. Hunter, *A Statistical Account of Assam*, vol. II, reprinted, Delhi, 1972, p.225.

16. W. J. Allen, op. cit., pp. 50-51.

17. W. W. Hunter, op. cit., pp. 223-224 mentions thirteen varieties of rice cultivated in the hills.

18. A. J. M. Mills, op. cit. p.4; W. J. Allen, op. cit., p. 50.

19. W. J. Allen, op. cit.,p. 44.

20. W. W. Hunter, op. cit., p.236.

21. Annual Report of the Khasi and Jaintia Hills in *Assam Annual Report 1877-1878*, p. 5.

22. For details of the limestone trade refer to Rita D. Dkhar, 'The Inglis and Company and the Lime Trade in the Khasi Hills,' unpublished Ph D. thesis, North Eastern Hill University, 1987; also refer to J. B. Bhattachajee, *Trade and Colony, The British Colonisation of North East India*, North East India History Association, Shillong, 2000.

23. David R. Syiemlieh, *British Administration in Meghalaya: Policy and Pattern*, Heritage Publishers, New Delhi, 1989, pp.52, 100-101.

24. D. R. Syiemlieh, 'Technology and Socio-Economic Linkages of the Khasi-Jaintia in Pre-Colonial Times,' in M. Momin and Cecile Mawlong (eds.), *Society and Economy in North East India,* vol. I, Regency Publications, New Delhi, 2004,pp. 21-34. A longer article of the subject of iron was earlier published; see D. R. Syiemlieh, 'Khasi Iron Culture and Iron Trade with Sylhet in the Late Eighteenth and Early Twentieth Centuries,' *Proceedings of the North east India History Association*, Eighth session, Kohima, pp.242-250.The most recent and detailed study on Khasi iron is researched by Pawel Prokop and Ireneusz, 'Two thousand years of iron smelting in the Khasi hills, Meghalaya, North East India,' *Current Science,* vol.104, No. 6, 25 March 2013, pp. 761-768.

25. W. Cracroft, 'Smelting of Iron in the Kasya Hills,' *Journal of the Asiatic Society of Bengal*, vol. I, 1832, pp. 150-151; a detailed study of the extraction of the ore, the labour involved and the wages paid to the workers is given in Lt. Yule, 'Notes on the Iron of the Kasia hills,' *Journal of the Asiatic Society of Bengal,* No. 129, 1842, pp. 854-855. Earlier H. Walters wrote a short note on the Khasi iron industry in 'Journey across the Pandua Hills near Sylhet in Bengal,' *Asiatic Researches*, vol. xvii, 1832, p. 505. Walters, Cracroft and Joseph Hooker in *Himalayan Journal*, vol. II, reprinted, Today and Tomorrow Printers and Publishers, New Delhi, pp. 306, have sketches of iron smelting furnaces in the Khasi hills.

26. Joseph Hooker, op. cit., pp. 292-293.

27. Robert Lindsay, *Anecdotes of an Indian Life*, with an introduction by David R. Syiemlieh, NEHU Publication, Shillong,1997,pp.37-38.

28. Captain Jones, 'Some Particulars regarding the mineral productions of Bengal,' *Gleanings in Science*, vol. I, 1829, p. 284.

29. Lt.Yule, op. cit., p.856.

30. W. W. Hunter, op. cit., p. 235.

31. W. J. Allen, op. cit., p. 48.

32. *Ibid*.

33. W. W. Hunter, op. cit., p, 235.

34. W. W. Hunter, op. cit., p.235.

35. W. W. Hunter, op. cit., p 241.

36. J. C. Thornton makes mention of these traders in *Memories of Seven Campaigns*, Westminster, 1895, p. 108.

37. *Ibid.*, p 108.

39. Hamlet Bareh, *The History and Culture of the Khasi People*, revised and enlarged edition, Spectrum Publications, Guwahati, 1985, p. 436.

39. W. W. Hunter, op., cit., pp. 131-132.

40. B. C. Allen, op. cit., pp. 92-93.

41. P. R. T. Gurdon, *The Khasis*, reprinted Low Price Publications, Delhi, 1996, p.190.

42. *Ibid*; Shobhan N. Lamare, *The Jaintias: Studies in Society and Change*, Regency Publications, New Delhi, 2005, p.136.

43. See Imdad Hussain, op. cit., pp. 35, 38, 88-89, for a short account with photographs of the first motor vehicles on the Cherra, Shillong, Gauhati roads.

6

Colonialism and Syiemship Succession: A Study of Cherra State (1901-1902)

Introduction

Each of the fifteen Khasi *himas* governed by *Syiems* have clans (*kur*) from which *Syiemship* succession takes place. Two main lines of succession are in practice. A deceased *Syiem* is succeeded by the eldest of the surviving brothers; failing such brothers by the eldest of his sister's sons. The succession rights apply only to those members of the family from the female line. There is a conventional preference to succession which goes down to nephews and grand-nephews. A female may become *Syiem* should there be no male heirs; she in turn would be succeeded by her eldest son.[1]

The method of election of *Syiems* differs from state to state. There being usually more than one candidate, an electoral *dorbar* constituted by *Lyngdohs*, *Basans*, *Lyngskors*, and in more recent times *Myntris* and *Sirdars*, with the founding clans of the state exercise authority to elect or reject candidates. In certain states all the male resident elect the *Syiem*,[2] while in certain others the decision of the electoral *dorbar* must be ratified by the people.[3] *Syiemship* is not hereditary but elective, and not necessarily from one family branch of the *Syiem Kur*. A *Syiem* rules over the subjects of the state. He has no territorial powers. His powers are limited, for in the evolution of *Syiemship* precaution was taken not to allow *Syiems* from becoming autocratic rulers. The founding clans play an important role in the Khasi states, not only having the privilege of electing the *Syiem* but also being entrusted with the government of the state with other officials through the *Dorbar Hima*.[4]

Soon after the British took control over the Khasi Hills and began to exercise their paramountcy, it became necessary in formalize succession to *Syiemships* through the Principal Assistant Commissioner in Cherrapunji. Where a claimant appeared in the dependent states and represented himself to be the heir of a deceased *Syiem*, he was required to state his right. When the officer was satisfied on the point, a proclamation was issued to all the inhabitants desiring them to state whether they accepted the right of the claimant. If objections were raised the people were called to vote whether they would or would not have the claimant. If the candidate obtained a majority of the votes he was considered duly elected, but if not, the people were allowed to elect any other member of the late *Syiem's* family who may be eligible according to the customs of the state. The Principal Assistant Commissioner would then confirm the candidate unless there were personal or political objections against him.[5] W. J. Allen in his report of 1858 suggested to the Government that succession to the four principal dependent states of Mylliem, Maharam, Mariaw and Nongkhlaw should be confirmed by Government, and further suggested that each of the *Syiems* be made to execute an *ikrarnamah* and that a *sanad* of appointment be given to them by Government.[6] Allen was of the opinion that:[7]

> This public and ceremonial recognition of these dependent Chiefs by the Government should probably be found very useful, inasmuch as it will ensure the respect and ready obedience of their subordinates, and this direct and visible responsibility to the paramount power will render these chiefs more anxious and careful to exercise their delegated functions with fidelity and discretion.

Of the five principal semi-dependent states it had been the practice to report to Government only the succession in Cherra. It was thought expedient that all the five states of Cherra, Khyrim, Nongstoin, Langrin and Nongspung should be reported to and receive the formal sanction of Government. Each *Syiem* should be required to present a *huzzur* and to receive a *khilat* from Government.[8] These suggestions of Allen were accepted by Government. From 1859, each Khasi *Syiem* was required to execute an agreement on his succession and in return he was conferred a *sanad*.[9]

Though Allen reported that there had been no reports of disputed succession,[10] Government was required to settle a number of succession disputes, the most serious being the dispute in Cherra State. Apparently, this was because while previously the Khasi states were sovereign severally and settled disputes according to the prevalent customs, with the Government of the East India Company and later the Crown assuming paramount power, disputed succession had to be settled by Government. British paramountcy encouraged claimants to *Syiemship* succession appealing to Government thereby circumventing the authority of the *dorbars*.

Succession in Cherra

Prior to British control over the Khasi states the method of election to *Syiemship* in Cherra state was that on the death of a *Syiem* the *Bakhraws* or elders of the 12 founding clans known as the *Khadarkur* would convene a dorbar and elect a successor. All power of election was vested in the *Khadarkur*.[11] Electing a *Syiem* by manhood suffrage did not exist till it was introduced by the British in cases of disputed succession.[12]

The decision of the *Khadarkur* was subsequently made known throughout the state by village elders.

A dispute arose in Cherra State on the death of Suba Singh, *Syiem*, on 5 June, 1856. His eldest nephew Ram Singh represented to the Principal Assistant Commissioner that he had succeeded his uncle. This was in turn reported to the Supreme Government, and the Governor General-in-Council sanctioned the succession. Ram Singh, however, had not been elected according to the custom of the state by the heads of the twelve clans. In consequence a great many of the influential people refused to acknowledge him as *Syiem*. Ram Singh was not permitted to perform the cremation rites of Suba Singh.[13] By Cherra custom, cremation of the late *Syiem* would confirm the election of the succeeding *Syiem*. The *Khadarkur* requested to elect a *Syiem*. It was thought inexpedient to hold another election as Ram Singh was not personally disqualified, and another election would have aroused passions and might have been attended with untoward consequences in the critical period of July-August 1857.[14] The *Khadarkur* were willing to have Ram Singh as their *Syiem* after Allen had explained that the succession of Ram Singh would not be taken as a precedent, but that in all future successions the ancient and established usages of Cherra state would be strictly abided.[15] Suba Singh was cremated on 4 May 1857.

Syiem Ram Singh died on 23 April 1875 and Hajon Manik succeeded to the *Syiemship*. It appears that Hajon Manik did not receive a *sanad* on his succession but was given a *parwana* in reply to which he stated that he had succeeded Ram Singh as the right of Bor Singh, the eldest cousin of the deceased *Syiem* was set aside as he had converted to Christianity. He recorded that "the 12 clans and elders constitute the Durbar for nominating a successor to a deceased *Syiem*."[16] The then Deputy Commissioner, Colonel Biver assumed that on the occasion of a succession to the Cherra *Syiemship*, the consent was not only the heads of the twelve clans but the "representative subjects" was to be sought.[17] For unknown reasons the family of Ram Singh declined to hand over the body of the deceased Syiem for cremation by Hajon Manik. Though Government recognised Hajon Manik *Syiem* of Cherra, the right to *Syiemship* of Hajon Manik had not been confirmed by religion and custom of Cherra state. Consequently Hajon Manik was never considered *Syiem* by the people of Cherra.[18] There was no appeal against Hajon Manik's appointment. The Cherra people considered Hajon Manik as *Nongsynshar* - administrator.

Reports on Succession

Prior to 1878 Government favoured the view that succession was determined by election and that the limitations which were imposed upon popular election owed their origin, in part to considerations of expediency. The custom of hereditary succession was to some extent lost sight of during Bivar's term of office. *Syiemship* was awarded to the person who had received the largest number of votes from the whole population. This practice "was not, however, uniform or consistant."[19] Biver's successor, Colonel Clark made a report on *Syiemship* succession. He was of the view, "that the system of election has for years been understood to be the custom among the Khasias, and in accordance with such understanding, elections, where the right to succeed has been contested, have from time to time been upheld."[20] Clark suggested that where disputes arose, an election should only be held on the demand of the *dorbar*. It was preferred that the *dorbar* should be unanimous in its decision as "elections conduces towards division in the communities and to much bitter feeling.' [21]

Hajon Manik died on 25 May, 1901. Captain D. Herbert, the Deputy Commissioner Khasi and Jaintia Hills District reported that the people of Cherra had held three *dorbars* on 26, 29 and 30 May and by a large majority had nominated Roba Singh of the house of Ram Singh to be their *Syiem*.[22] A minority were in favour of Chandra Singh, a nephew of Hajon Manik. By Cherra custom Chandra Singh was disqualified as all the female members of his house and family had died which is considered a disqualification by the Khasis. He was further disqualified as he was of the house of Hajon Manik who was "no Syiem", as Ram Singh's cremation had not been performed. [23] Satisfied that Roba Singh had been nominated with fairness, the Deputy Commissioner recommended that the nomination of Roba Singh be accepted and that he be appointed *Syiem*.[24]

The Chief Commissioner of Assam, Sir Henry Cotton, doubted the manner in which Roba Singh had been nominated by the *dorbars*. He believed that Chandra Singh was 'the heir' as he was directly related to Hajon Manik. He instructed the Deputy Commissioner to convene the *Khadarkur* and to call them to state whether they would consent to the succession of Chandra Singh. He further ordered that if Chandra Singh's claims were not consented, an election by the inhabitants of Cherra state should be held.[25] Captain Herbert held a *dorbar* of the heads of the twelve clans on 9 July, 1901. There were by then three claimants. Apart from Roba Singh and Chandra Singh, one Bolo Singh also presented his claim. 8 of the 12 men consented to the election of Chandra Singh and 4 opposed his election: and stated that they desired Roba Singh. Herbert had interpreted the Chief Commissioner's orders of 18 June to mean that if the *dorbar* was not unanimous, an election by the people of the state should be held. He therefore called for an election on 14 August.[26] Herbert's views were that an election by the people was advisable as the objection to Chandra Singh was of a religious nature. He urged that W. J. Allen's remarks relating to the power of the *Khadarkur* to elect a *Syiem* should be read with, and need not conflict with more recent opinion to the effect that in cases of contested succession, the people should elect the *Syiem*. The *Khadarkur* being divided over the issue, and as the majority of the *dorbar* continued to support Chandra Singh against the wishes of the people, a feeling of distrust had arisen.[27]

When the Cherra succession case came before the Chief Commissioner, he reminded the Deputy Commissioner that an election by the people of Cherra was improper as it had not been called for by the *dorbar*, as Colonel Clark had suggested earlier, and which had been approved by the then Chief Commissioner, Sir Steuert Bayley. [28] The procedure of popular election, they resolved, "is contrary to the custom of the state," and that "it would be a dangerous precedent to proceed to popular election on the present occasion." The orders for an election by the people was set aside and Chandra Singh was appointed *Syiem*.[29]

Official Intervention

Chandra Singh was not *Syiem* for long. The great many of the Cherra people refused to acknowledge his authority. He did not have the tact to reconcile his opponents or the power to coerce the opposition. While Chandra Singh complained of lawless conduct on the part of Roba Singh and his adherents, complaints were also received of alleged acts of oppression by the *Syiem*.[30] On reports of violence in Cherrapunji, the Deputy Commissioner accompanied two companies of the 43rd Gurkha Rifles to the town on 15 December 1901. The order of appointment of the *Syiem* was withdrawn on 30 December.[31]

What set the process of election once again were the reports of the Deputy Commissioner to the Chief Commissioner who in turn kept the Governor-General and the Government of India, Foreign Department informed of the developments in Cherra. In all probability the appeal of Roba Singh to the Government of India[32] and the memorials of U Titor, Myntri and others[33] and the people of Cherra state[34] influenced Government to review afresh the claims of Chandra Singh and Roba Singh. The memorial of U Titor, *Myntri* suggested that a grand dorbar, be assembled by the heads of the twelve clans in which the elders of the Cherra villages should also be present. In the meantime Captain Herbert was requested by the Viceroy that he should proceed to Calcutta.[35] One may reasonably assume that Herbert was able to personally inform the Viceroy, Lord Curzon what had occurred in Cherra and the reasons for Chandra Singh's unpopularity. The Viceroy taking very careful consideration of the Chief Commissioner's minutes on the Cherra case was reluctantly compelled to come to the conclusion that the recognition of Chandra Singh as *Syiem* of Cherra could not be confirmed, and that an election by the inhabitants of Cherra state should be held as was ordered by the Deputy Commissioner on the 9 July 1901.[36]

The election was fixed for 3 April 1902. Chandra Singh was ordered on 17 March to hand over charge of the state to the *Myntris*. He regretted he could not do so, neither could he be present in Cherrapunji as he was then in Shillong. He requested the postponement of the election as he wished to appeal against the order of Government.[37] Certain *Myntris* of the *Khadarkur* too wanted a postponement of the election for two months.[38] They were not willing to take charge of Cherra state and would appeal against the decision to hold a popular election.[39]

The election was held on the date notified. There was no actual contest as Chandra Singh withdrew and declined to stand as a candidate by general election. He and the 8 *Myntris* of the 12 clans retired altogether from Cherrapunji and were in Shillong at the time of the election.[40] Serious disturbances broke out in Cherra. It was alleged by the Deputy Commissioner that the adherents of Chandra Singh fired on people going to the election and thus intimidated the majority of the voters from attending it. One of Roba Singh's men was shot dead. It is reported that Roba Singh's adherents had also taken to arms as three of Chandra Singh's men received gunshot wounds.[41] It appears that in the election 747 voters were present, but apparently there was no record of voters. The Deputy Commissioner enquired whether the people wished to vote for Roba Singh or any other candidate and as all seemed to be in favour of Roba Singh, he counted their number and declared Roba Singh to be duly elected[42].

Roba Singh was confirmed as *Syiem* of Cherra by the Government of India Foreign Department[43], which in turn was an authorisation for the Chief Commissioner to issue a *sanad*.[44]

Chandra Singh's *sanad* was cancelled. Convinced that he still had a legitimate claim Chandra Singh addressed two memorials to the Secretary of State for India for a reconsideration of the election proceedings just concluded.[45] Chandra Singh's petition to Government ultimately reached the House of Commons.He was informed that the Secretary of State had accepted the decision of the Government of India and was not willing to interfere on his behalf.[46] Roba Singh's election as *Syiem* was confirmed by custom when on 6 March 1908, he cremated the dead body of Ram Singh.[47] Hajon Manik's cremation was done by his family members.

Conclusion

The Cherra *Syiemship* succession case 1901-1902 reveals that with the best of intentions the various levels of Government had to review the succession three times over. While the Deputy Commissioner, Captain D. Herbert and the Government of India, Foreign Department were of the opinion that Roba Singh had more rights than Chandra Singh, the Chief Commissioner, Sir Henry Cotton upheld the claim of the latter. The one fact that emerges from the succession to *Syiemship* in Cherra State is that it was never sufficiently and clearly defined to be indisputable and that the accepted method varied from time to time. It is apparent that Government was abiding by principles and guidelines in not accepting the decision of the people of Cherra state in the first election, as the three *dorbars* had not been called for by the *Khadarkur*. In the second election, 8 of the *Khadarkur* who were in favour of Chandra Singh were apparently personally interested in their candidate. The people of Cherra thus, did not accept their decision, nor did they abide by Government's recognition of Chandra Singh as *Syiem*. The third election by the adult male population was ordered by the Governor- General, and not by the electoral *dorbar* Colonel Clark had earlier suggested. This change in policy is clearly explained in the correspondence of the Secretary to the Chief Commissioner, to the Secretary, Government of India, Foreign Department in which he explained: "In the present case (popular election) it has undoubtedly served to prevent the succession of a man whose authority could on religious grounds never have been recognized by the great majority of the people."[48]

Notes and References

*David R. Syiemlieh, 'Colonialism and Syiemship Succession: A Study of Cherra State (1901-1902), *Proceedings of the North East India History Association*, Third session, Imphal, 1982, pp.147-157.

1. P.R.T. Gurdon, *The Khasis*, reproduced, Low Price Publications, Delhi, 1996, p. 71.

2. As in Langrin and Nobosohphoh; *Ibid.*, p. 74.

3. As in Mawiong, Mawsynram, Bhowal and Malaisohmat.

4. There is no exact English equivalent for the Khasi term *Hima*. The closest meaning would be 'State'.

5. W. J. Allen, *Report on the Administration of the Cossyah and Jynteah Hill Territory 1858*, Calcutta, (reprinted) 1903.

6.   *Ibid.*, p.79.

7.   *Ibid.*

8.   *Ibid.*, p. 77.

9.   C.U. Aitchison (comp.), *A Collection of Treaties; Engagements and Sanads Relating to India and Neighbouring Countries*, Calcutta, 1931, p. 84. It was further clarified: "The formal investiture of a chief should, if possible, be performed by a British Officer. Such a course may not always be practicable, but I am to observe that the succession to a Native State is invalid until it received in some form the sanction of the British authorities. Subsequently an ad interim and unauthorised ceremony carried out by the people of a State cannot be recognised, although the wishes of the ruling family and the leading persons in the State would naturally in all cases receive full-consideration." William Lee Warner quoting the *Gazette of India,* 22 August, 1891, *The Native States of India*, Macmillan and Co., London, 1910, p. 322.

10. Allen, op. cit, p. 30.

11. *Ibid.*, p. 82.

12. Assam Secretariat Records, Foreign Department Proceedings A (hereinafter ASR, FDP A) May, 1902, No. 38, Klur Singh, *Syiem* of Khyrim to Secretary, Chief Commissioner, Assam.

13. W. J. Allen, op. cit, p.82.

14. *Ibid.*, pp.82-83.

15. *Ibid.*

16. ASR, FDP A, February 1902, No 56, Translation of the report dated15 August, 1878, of U Hajon Manik, Syiem of Cherra in Durbar.

17. ASR, FDP A, September 1902, No. 20, Secy, to Chief Commissioner to Secy. Government of India, Foreign Department, dated 9 June, 1902.

18. ASR, FDP A, November 1901, No.5, Captain D. Herbert to Secretary to Chief Commissioner Assam.

19. ASR, FDP A, March 1902. No. 15, Secretary to Chief Secretary to Secretary Government of India, Foreign Department, 28 February 1902.

20. ASR, FDP A, November 1878, No.1, Colonel Clark to Secretary to Chief Commissioner, 15 October 1878.

21. *Ibid.*

22. ASR, FDP A, November 1901 No.2, Captain D. Herbert to Secretary to Chief Commissioner, 1 June 1901.

23. *Ibid.* ASR, FDP A, November 1901, No.5, Captain D. Herbert to Secretary to Chief Commissioner, 19 July 1901.

24. ASR, FDP A, November 1901, No.2.

25. ASR, FDP A, November, 1901 No.3, Secretary to Chief Commissioner to Captain D. Herbert, 18 June 1901.

26. ASR, FDP A, November 1901, No 4, Captain Herbert to Secretary to Chief Commissioner, 9 July 1901, and ASR, FDP A, November 1901, No.5, Captain D. Herbert to Secretary to Chief Commissioner, 19 July 1901.

27. *Ibid.*

28. ASR, FDP A, November, 1901, No. 10. Resolution on the succession to the Chiefship of the semi independent state of Cherra on the death of late Syiem U Hajon Manik, 21 July 1901.

29. *Ibid.*

30. ASR, FDP A, February 1902, No 49, Secretary to Chief Commissioner to Secretary Foreign Department, 16 December 1901.

31. *Ibid.*

32. C. U. Aitchison, op. cit., p.87.

33. ASR, FDP A., March 1902, No.7, Memorial by U Tibor Myntri and others to the Chief Commissioner of Assam, 18 January 1902.

34. ASR,FDP A, March 1902 No.2, Memorial sent by the people of Cherra state to the Viceroy, 20 October, 1901.

35. ASR, FDP A, March 1902,No. 10, Telegram to Chief Commissioner Assam, 25 January 1902.

36. ASR, FDP A, March 1902 No. 14, Secretary Government of India, Foreign Department to Chief Commissioner, 11 February, 1902.

37. ASR, FDP A, May 1902, No. 32, Chandra Singh to Deputy Commissioner, 24 March 1902.

38. ASR, FDP A, May 1902, No. 35, U Tibor to Secretary Government of India Foreign Department.

39. ASR, FDP A, May 1902, No. 34, Translation of a petition from Khasi to English from U Titor and other *Myntris* to Deputy Commissioner, 24 March 1902.

40. ASR, FDP A, May 1902, No. 26 Secretary to Chief Commissioner to Secretary Government of India, Foreign Department, 7 April, 1902.

41. *Ibid.*

42. *Ibid.*

43. ASR, FDP A, May 1902, No. 43, Telegram from Government of India, Foreign Department, 19 April, 1902 communicated by Secretary to Chief Commissioner to Deputy Commissioner, 21 April 1902.

44. ASR, FDP A, May 1902, Officiating Secretary to Chief Commissioner. to Deputy Commissioner, 9 June 1902.

45. ASR, FDP A, September 1902, Nos. 10 and 18, Memorials of Chandra Singh to Secretary of State, 7 May 1902 and 29 April, 1902.

46. ASR, FDP A, December 1902, No.6, Under Secretary Government of India Foreign Department to Chief Commissioner Assam, 13 November, 1902.

47. *U Hynniew Trep*, 18 December 1981.

48. ASR, FDP A, September 1902, Secretary to Chief Commissioner to Secretary, Government of India, Foreign Department, 9 June 1902.

7

British Policy Towards the Khasi States

Introduction

For about sixty years after the East India Company secured the Dewani of Bengal its relations with the Khasi states was minimal. The Company's interest in the new Sylhet frontier was one in which the limestone trade from the hills could be encouraged to the benefit of British traders; to check the frequent raids of the hillmen into the plains and to stake a firm claim over the foot-hills which the Khasi chiefs also claimed as forming portion of their *himas*. If the British had no political interest in the hills, they were initially concerned about the activities of Greek and French traders in Sylhet, fearing that the Greeks planned to establish a colony outside the Company's territories within the Khasi Hills, and that the French acting through one M. Dechampigny who had attempted to establish political relations with the Khasis might establish a protectorate on the frontier.[1] The perpetuation of these interests and the successful campaign against Ganga Singh, *Syiem* of Shella, and his allies who had challenged British control of the foot-hills resulted in the Company adopting a number of measures by which traders were prohibited from entering into the hills and stern steps were taken to control Khasi raids into Sylhet.[2]

Early Political Control

Until the close of the eighteenth century British relations with the Khasi states practically did not operate other than with those states that abutted on Sylhet. No European appears to have travelled through the hills, which explains their ignorance of the Khasi states' structure, their government and the way of life of the people. The Company was then following a policy of non-intervention towards the entire North-

East. Such a policy, however, could not continue for long. The Burmese occupation of Assam and the threat this occupation posed to the British in Bengal meant the abandonment of the policy to one of a much closer political contact-one of intervention in the affairs of Assam, Cachar, Manipur, Jaintia, and the Khasi states. The Khasi states were the little affected by the Anglo-Burmese War of 1824-26. Its aftermath, however, was to have a telling effect on the status of the Khasi states. David Scott, the Agent to the Governor-General, North-East Frontier who had resided in the hills for a time during the parleys with the Jaintia Raja was delighted by the climate of the hills. Scott brought it to the notice of the Government that locations in the Khasi Hills could be developed for sanatoria stations for Europeans in the Lower Provinces. Scott constructed a house in Nongkhlaw "to eat the Europe air."[3] Scott was also anxious to construct a road through the Khasi states to connect Sylhet and Assam as the one started through the Jaintia Hills had been discontinued. By a very tactful move[4] Scott was able to secure from Tirot Sing, *Syiem* of Nongkhlaw, the necessary permission for the construction of a road through his state and for a sanatorium at Nongkhlaw. Dewan Singh, *Syiem* of Cherra likewise gave his consent to these facilities.[5]

It is not necessary here to give the origins of the Anglo-Khasi War that broke out with the "Nongkhlaw massacre" of 4 April 1829. Suffice it to say that the Khasis, both chiefs and people were worried by the manner in which the British had penetrated into their hills, what with the arrival of soldiers and convicts from Assam and Sylhet to construct the road, the start of a sanatorium at Nongkhlaw and later at Cherrapunjee and the British control of the northern and southern foot-hills. The official apology for the outbreak has been ascribed to "the false and foolish speech of a Bengalee Chupprassee" who in a dispute with the Khasis, "had threatened them with his master's vengeance and had plainly told them that it entered into his master's plans to subject them to taxation, the same as the inhabitants of the plains".[6] All this created a psychosis of fear the expression of which was to involve practically all the Khasi states in a serious conflict with the British which was eventually put down by British forces with the surrender of Tirot Singh on 13 January 1833.

Those states that had actively participated in the struggle either lost a portion of their territories; were made to pay fines; or a combination of the two measures. No *hima* was annexed into the British dominion of India other than three small villages of Mawsmai, Mawmluh and Sohbar, the first two by conquest and the third by treaty.[7] Scott personally believed that it was more expedient to impose fines than collect tribute because the cost of collection would outweigh the receipts apart from the risks involved in its collection.[8] Francis Jenkins who succeeded William Cracroft and T.C. Robertson as the Governor General's Agent agreed in principle that to desist from levying these fines in part or in whole would be to expose the hillmen to commit further offences.[9] Government at Calcutta while agreeing with these measures cautioned that since the "indiscreet use of which might severely press upon the resources of. communities and injuriously affect the reputation of the British Government, the exercise of it ought to be kept within well defined limits".[10] In time the British recognized twenty-five Khasi states as semi-independent and dependent states. Cherra, Khyrim, Nongstoin, Langrin and Nongspung were referred to as semi-independent states as they had never been actually coerced by a British force. The remaining twenty states recognized as dependent states were those that had been restored to their chiefs, or in the case of the five *Sirdarships* were those states that were created by the British. [11]

The *Syiems* of the semi-independent states were permitted to exercise, with the aid of the *durbars* and elders, sole criminal and civil jurisdiction in their respective states over their own subjects on matters pertaining exclusively to them. The dependent chiefs were permitted to investigate and decide all civil and criminal cases in which the parties were their own people with the exception of cases of murder, homicide and accidental death which were reported to the Cherra authorities for their decision. Both categories of states were bound to take to the Cherra court cases in which British subjects and the inhabitants of more than one state were concerned. [12] The Khasi states came within the jurisdiction of the Political Agency at Cherrapunji when it was instituted on 11 February 1835 under Captain Lister, the Political Agent and his Assistant and son-in-law, Henry Inglis.

British paramountcy was further extended over the Khasi states by formalizing succession of chieftainship. Though the British recognized hereditary and elective succession as was in vogue, formal recognition of succession in the dependent states only followed after the Principal Assistant Commissioner had verified the claimants, rights and after no objection came from a state's population. [13] If objections were raised the people were called to vote whether they would or would not have the claimant as chief. If the claimant obtained a majority vote in his favour he was considered duly elected. Should the claimant have failed to secure the confidence of the electorate, another election was held to elect any other person eligible for the office. The Principal Assistant Commissioner would then confirm the candidate's succession unless there were personal or political objections against him. [14] *Syiems, Lyngdohs, Wahadadars* and *Sirdars* were made to realize that their appointments came from the British Government and that they could be removed if they did not cooperate with the authorities. [15]

The foregoing account explains that the British administered the entire Khasi Hills, except for the Cherrapunji station indirectly through the traditional leaders. A distribution of power and authority was arrived at without the British having to assume the entire responsibility of administration in what has been called 'indirect administration.' The number of British officials was small necessitating collaboration from the Khasi chiefs. [16]

Extension of Paramountcy

Even after the changes in administration that came with Mills' report it was found by the British that this relationship with the various Khasi chiefs was not formally defined. Lord Dalhousie therefore minuted: "that the paramount and direct authority of the British Government over the whole assembly of these petty chieftainships as well as over the remainder of the territory comprised within the Agency should be asserted and proclaimed in legal form," with the view not to extend the interference of the British authorities in the affairs of the Khasi chiefs or to alter the kind of degree of subjection in which they were placed, but to legalize the power which had been hitherto exercised by the Political Agent. [17] It was left to W. J. Allen of the Board of Revenue to make suggestions for Government to consider. Succession to the five semi-independent and the four principal dependent states of Mylliem, Maharam, Myriaw and Nongkhlaw should be reported to the Government for its confirmation. Each succeeding *Syiem* of these nine states should be required to present a *huzzur* to, and recieve a *khilut* from the Government and each *Syiem* should execute *ikrarnamah* and receive a *sanad* of appointment from the Government.[18] Allen did not think that it would be necessary that these *ikrarnamah* should contain any minute article of agreement. It would suffice for the *Syiems* to promise to govern their states according to its ancient and established usages, to reside in the state, to keep the people contented and satisfied and to obey any orders that might be given to them by the paramount power. The stipulation regarding residence was considered necessary because the state of Mylliem had been in deplorable condition owing to the absence of its *Syiem*, Hazar Singh, who had been for many years residing with his wife's family at Cherrapunji.[19] Succession to the remaining states should be reported to the Governor-General's Agent who might be empowered to grant *sanads* of succession.[20]

The assumption of power over India by the Crown in 1858 brought in more definite and formal relationship between the Khasi states and the British. In 1859, following Allen's report it was decided to require the execution of a general agreement by each of the Khasi chiefs on their succession with the district's Deputy Commissioner who was also to function as Government's Political Officer in relation with the states. Periodic changes in the terms of the agreements taken from the chiefs were made between 1864 and 1875 when it was decided that appointments of chiefs should rest with the Governor-General in the case of larger Khasi states and with the Chief Commissioner of Assam in the case of the smaller states. In November 1875 the system of signing agreements was abolished altogether. Instead it was decided that recognition given by Government of a succession to a Khasi state should take the form of a *sanad* conferred upon them instead of an agreement taken from them. After 1877 the sanctioning authority became the Chief Commissioner in the case of *Syiems* and the Deputy Commissioner in the case of *Lyngdohs, Sirdars,* and *Wahadadars.*[21] This system continued till 1912. In June 1910 a change was suggested by the Lieutenant -Governor of Eastern Bengal and Assam that since the Khasi states were "small and unimportant" and since the status of *Syiems* was not such as to warrant the issue to them of *sanads* by the head of the province, all future *sanads* should be given by the Commissioner of the Surma Valley and Hill Districts Division.[22] This came into effect two years later.

In the new relationship the British Government could intervene in the internal-affairs of the states. Intervention took various forms, over maladministration; over discontent of a state's subjects with their chiefs; over questions of law and order and in most cases; over disputed succession to chieftainship. Often advantage was taken of internal dissensions and inter -village dispute within a Khasi state to allow villages to secede from states and become British villages.[23] The Government intervened in Sheila when its *Wahadadars* were found to be maladministering that state. Suggestions were put forward that instead of four *Wahadadars* there should be one; that the chief should be a *Syiem* and that the state should be taken over by the British Government. When these options were put to the people, their reply was for an adherence to their old custom and; therefore, four *Wahadadars* continued to administer Shella.[24] *Syiems* were deposed for various faults. Mudon Singh, Syiem of Nongstoin was removed in 1893 for deliberately hushing up three serious cases of murder.[25] Symburai, *Syiem* of Mawsynram was convicted in 1904 for his involvement in two murder cases and transported for life to the Andamans where he remained till 1926.[26] When ill health made it difficult for Klur Singh, *Syiem* of Khyrim attend to public affairs he applied in August 1902 that his two nephews be entrusted with the management of his state. Accepting the *Syiem's*, inability to govern but refusing the handing over of administration to the two nephews, the Government appointed *Babu* Hajom Kissor Singh, a head clerk in the subdivision office at Jowai, an "intelligent, active, tactful and absolutely reliable person" to be the *Dewan* of Khyrim for one year.[27]

Syiemship Succession

But by far the most perplexing problem for the British administration was the issue of *Syiemship* succession. Deputy Commissioner, H.S. Bivar was of the opinion that *Syiems* were elected. The Chief Commissioner, Colonel Keatinge relying on the Deputy Commissioner's local knowledge and long residence in the hills accepted his views. *Syiemships* were therefore awarded to the person who had received the largest number of votes from a state's adult population. But Bivar's views were wrong and he had misguided Keatinge. Every election which had been made on the basis of election since the formation of the Assam Chief Commissionership had been appealed against. Keatinge directed that full enquiries be made for purposes of ascertaining more explicitly what ground existed for the adoption of the principle of election.[28] Exhaustive enquiries were accordingly made by Bivar's successor, W.S. Clarke. The conclusions he arrived at was that the office of *Syiem* was not elective but *hereditary* going in regular succession to the *Syiem's* nearest male relation from the female line; that the candidates must be blood relations of a *Syiem*, and that on the death of a *Syiem* the next in regular line came forward and proclaimed himself *Syiem* with the consent of the family and the durbar. Clarke's opinion was that election conduces towards division in the communities and much "bitter feeling". He proposed that an election:[29]

> Should only be held on the demand of the durbar who should, on the death of a Siem, be summoned, with the members of the Siem's family, and be called to nominate the new Siem. The fact of the nomination should be formally recorded, and the person nominated should no reason against his appointment existing be forthwith proclaimed Siem.

The Chief Commissioner, Stewart Bayley approved this proposal which was then and later accepted as an authoritative rule on the subject.[30]

The first test to this policy occurred in a disputed succession in Nongstoin. On the deposition of Mudon Singh, an election was held on 13 April 1894 to elect a new *Syiem*. Two candidates presented themselves, Rabon, cousin of the late *Syiem* and Indro. Rabon was elected *Syiem* by a majority of 572 votes, against 559 and was confirmed as *Syiem*. Rabon died in 1897 and his brother Mon was appointed to the office without an election. Sowon, brother of Mudon Singh claimed that he too was eligible for *Syiemship*, but J. C. Arbuthnott, the Deputy Commissioner would not allow the family of Mudon to have any influence in the state. His case was summarily dismissed. Arbuthnott's comment is worth quoting.[31]

> I am altogether opposed to the election system. It only leads to bribery and generates party feeling, and is adverse to the interests of the people. I also believe that elections are an innovation consequent on our rule and contrary to Khasi custom.

Another, and far more complex succession case occurred in Cherra. On the death of Hajon Manick on 25 May 1901 the people of that state had in *dorbar* nominated Roba Singh to be their *Syiem*. A minority were in favour of Chandra Singh. First Roba Singh succeeded, then from orders of the Chief Commissioner the right of Roba Singh was set aside for the rival claimant. Yet a third succession was to take place on the intervention of Lord Curzon, the Governor-General. An election was held on 3 April 1902 that went in favour of Roba again.[32] Petitions were made to the Government by the *Syiems* of Khyrim, Nongkhlaw, Nongstoin and Mylliem against the principle followed in the Cherra case.[33] Chandra Singh's petition to Government ultimately reached the House of Commons.[34]Fearing that more problems might arise were the system used in Cherra be applied to the other *Syiemships*, the Government of India ordered an enquiry into the tradition of appointment to *Syiemships* in the Khasi Hills by D. Herbert, the Deputy Commissioner.[35]

Herbert's report in January 1903 showed that no general rule would suit the circumstances of all the *Syiemships*. He found the people generally adverse to popular election and anxious for a restoration of the old customs of nomination in the first instance by certain constitutional electors. He recommended that in the event of a vacancy occurring in a *Syiemship* the Deputy Commissioner should call on the electors to nominate a *Syiem*, and that only in the event of objections to such nominations being lodged, and in certain cases if the electors were unanimous should resort be taken to record the votes of the electors.[36] The Chief Commissioner, Sir J.B. Fuller generally accepted Herbert's changes but made it clear that in all cases of dispute over succession where the electors were equally divided, the Government should decide as to which candidate had the strongest claim. "Of course," he added, "Government should. always reserve to itself a right of refusing to accept a person as Siem whose appointment would be undesirable for any serious reason." The Government had exercised in the past the power of setting aside a nominee and it should retain this authority in future, he added.[37]

Such a policy of interference and disregard to the customs of the Khasi states and the low position the chiefs held in the estimation of British administrators was perhaps due to the size and population of the states. Malaisohmat had a population in 1871-72 of only 299. Few states had populations of over a thousand subjects. Even the largest of the states, Khyrim, had a population of 20,504.[38] Though the Calcutta High Court had ruled in 1884 that the Khasi chiefs were ruling chiefs, recognized as such by the Government, "by some curious mischance," this ruling was lost sight of and in the absence of guidance the chiefs were, till the discovery of the ruling in 1933, at various times treated with degrees of respect ranging from that due a ruling chief to what is due to a village headman.[39]

Changing Dimension of Relations

Though political activity can hardly be said to have existed in the Khasi states, their chiefs and subjects were showing signs of increased political awareness from the 1920s. The expression of this is evident in their various statements, memorials and the like that they prepared and submitted to various authorities in 1921, 1928-29 and in 1932 in a conference of Khasi states. The gist of these showed their loyalty to the British Government while at the same time a criticism of British policy towards the states. The chiefs particularly wanted a reconsideration of their status.[40] It may be not out of place to show that so wide were the powers of the Deputy Commissioner that one of them in 1910 even considered the states as part of British India![41] In the wake of constitutional changes in India towards the end of British rule, Khasi chiefs were anxious to know what their future would be. Lord Irwin, the Viceroy, in a reply to an address presented by the people of the Khasi Hills made it clear that: "Whatever the constitutional developments may be, I have no doubt that the rights and privileges of the Siems will be safeguarded, and that, so far as may be practicable, steps will be taken to preserve the national individuality of the Khasi race."[42]

In line with the general policy of the British in giving the Indian states a political role, and after the "discovery" of the Calcutta High Court ruling of 1884 that the Khasi chiefs were ruling chiefs,[43] Lord Linlithgow the Governor-General advised the Khasi states to federate.[44] In 1934 was established the Federation of Khasi States which despite the enthusiasm for such an organization was all but defunct a few years later. The hopes that the Indian princes would cooperate in the proposed Federation envisaged in the Act of 1935 and the failure of this plan must have in a way resulted in the Khasi states, like many other Indian states, losing the political momentum that had aroused interest in their future.

By the late 1930s it was found that the Deputy Commissioner of the Khasi and Jaintia Hills as Political Officer of the Khasi states had tended to become immersed in his preoccupations as Deputy Commissioner of a British district to the neglect of the duties of an officer "who ought to be the trusted adviser of the Chiefs".[45] A suggestion had been put forward to Sir Robert Reid, the Governor of Assam to separate the two posts. Reid's own view on this was that the financial implications alone would have made it impossible to implement. He however felt that at least something should be done by way of emphasizing more decidedly the chief's rights and keeping clearly in mind the distinction between the Khasi states and British India as well as by selecting for the post of Political Officer "a man who could regain the lost confidence of the chiefs".[46]

The Federation of Khasi State was revived in 1946 shortly before the lapse of British paramountcy. This came in the wake of developments that took place with the transfer of power by the British Government. The Khasi states were left with the options of remaining independent; of joining one or the other of the two dominions and a third alternative of federating among themselves. The legacy of the *Raj's* relations with the Indian states posed a serious problem to the new Indian Dominion. In its attempt to get the Khasi states to join independent India much tact and some force had to be used, for the new frontier had suddenly given new importance to the Khasi states, situated as many were on East Pakistan's frontier with India.[47]

Notes and Reference

*David R. Syiemlieh, 'British Policy towards the Khasi States', J. B. Bhattacharjee (ed.) *Studies in the History of North-East India: Essays in Honour of Professor H. K. Barpujari*, North-Eastern Hill University Publications, Shillong, 1989, pp.187-196.

1. W.K. Firminger, *Sylhet District- Records*, vol. II, Shillong 1917, No. 116; B.C. Allen, *Assam District Gazetteer*, vol. II, Calcutta, 1905, pp, 52-53.

2. For details see P.N. Dutta, *Impact of the West on Khasis and Jaintias*, New Delhi, 1982, pp. 27-46; D.R.Syiemlieh, *British Administration in Meghalaya: Policy and Pattern*,Heritage Publishers, New Delhi,1989, pp, 16-24.

3. "Chirra Punji," *Asiatic Journal*, New Series, vol. XXI, 1836, p. 17; A White, *A Memoir of the Late David Scott*, Calcutta, 1832, p.37.

4. The annexation of Assam deprived many of the Khasi states of the nine *duars* (passes) they held in fief from the Ahom monarchy. Taking advantage of a disputed succession in Nongkhlaw, Scott announced that he would not grant the investiture of the Barduar controlled by Nongkhlaw to anyone whose authority was not fully recognized in the hills and the lowlands, and who was not prepared to grant the same facilities of communication as the Khasi enjoyed in Assam and Sylhet, On Scott's mediation the claims of Rajan Singh, a lad of five years and brother of the deceased *Syiem*, were set aside for Tirot Sing. See N.K. Barooah, *David Scott in North East India*, Munshiram Manoharlal,New Delhi, 1970, p. 179.

5. A. White, *A Memoir of the Late David Scott*, Calcutta, 1832, pp. 32 37; C.U. Aitchison, *A Collection of Treaties Engagements, and Sanads*, Vol. XII, Calcutta, 1931, pp. 122 23; "Chirra Punji," *Asiatic Journal*, Vol. XXI, 1836, p. 17.

6. A. White, *A Memoir of the Late David Scott*, Calcutta, 1832, pp. 41-42.

7. A.J.Mills, *Report on the Khasi and Jaintia Hills,* Calcutta,1853, reprinted, Shillong, 1901, p. 7.

8. National Archives of India, Foreign Political Consultations, 27 May 1834, No. 78.

9. *Ibid.*, 7 August 1834, No. 57.

10. *Ibid.*, 22 May 1834, No. 83.

11. W.J. Allen, *Report on the Administration of the Cossyah and Jynteah Hill Territory,* Calcutta,1858, reprinted, Shillong, 1903, p. 28; C.U. Aitchison; op. cit., p. 238; P.R.T. Gurdon, *The Khasis*, Delhi, reprint 1975, p. 75.

12. W.J. Allen, *Report on the Administration of the Cossyah and Jynteah Hill Territory*, pp. 26-28; A.J.M. Mills, *Report on the Khasi and Jaintia Hills*, pp. 67.

13. Mill's report on the Khasi and Jaintia Hills of 1853 had advocated for a complete change in the powers and functions of the British administration in the Khasi and Jaintia Hills. Following the promotion and retirement of Lister, the Political Agency at Cherrapunii ceased. C.K. Hudson took charge in April 1854 as Principal Assistant Commissioner, Khasi and Jaintia Hills District, attached to the Assam administration.

14. W.J. Allen, *Report on the Administration of the Cossyah and Jynteah Hill Territory*, pp, 28-29.

15. C.U. Aitchison, *A Collection of Treaties Engagements, and Sanads,* vol. XII, pp. 144,155-56.

16. David R. Syiemlieh, *British Administration in Meghalaya: Policy and Pattern*, pp.69-71.

17. A.J.M. Mills, *Report on the Khasi and Jaintia Hills*, p. 118.

18. W.J. Allen, *Report on the Administration of the Cossyah and Jynteah Hill Territory*, pp,77-79.

19. *Ibid.*, p.77.

20. *Ibid.*, p.80.

21. C.U. Aitchison, *A Collection of Treaties Engagements, and Sanads*, vol. XII, pp. 84-87.

22. Assam Secretariat Records, hereafter ASR, Political Proceedings, June 1910, Nos. 29-30.

23. By the turn of the nineteenth century the number of British villages had increased to 35.

24. ASR, Foreign Proceedings, August 1876, Nos.15-17, June 1887, No 5.

25. *Ibid.*, July 1903, p. 4. Hereinafter referred to as D. Herbert, *Report on Succession to Siemships in the Khasi Hills*.

26. *Assam Administration Report 1903-1904*, p. 5; C.U. Aitchison, *A Collection of Treaties Engagements and Sanads.*

27. ASR, Foreign Proceedings, October 1902, Nos. 13-15; October.1903, Nos. 1-7.

28. *Ibid.*, March 1902, No.15.

29. *Ibid.*, November 1878, No.1; *Report on Succession to Siemships in the Khasi Hills*, p, 2.

30. C.U. Aitchison, *A Collection of Treaties Engagements, and Sanads*, Vol. XII, pp, 84-87. *Ibid.*, No.2; D. Herbert, Report on Succession, p. 2.

31. D. Herbert, *Report on Succession to Siemships in the Khasi Hills*, p. 4.

32. For a detailed account of this succession, see D.R. Syiemlieh, *British Administration in Meghalaya,* op. cit., pp. 157-164; D.R. Syiemlieh, "Colonialism and Syiemship Succession: A Study of Cherra State (1901-1902)," *Proceedings of the North East Indian History Association*, Third Session, Imphal, 1982, Shillong, 1983, pp. 147-57.

33. D. Herbert, *Report on Succession to Siemships in the Khasi Hills*, Nos. 14-15; ASR, Foreign Proceedings, May 1902, No. 38.

34. David R. Syiemlieh, *British Administration in Meghalaya: Policy and Pattern*, pp.161-162; Hansard, Commons,4 series, vol.130,1904, pp.235-236; vol. 139, 1904,p. 967.

35. *Ibid.*, No. 14.

36. D. Herbert, *Report on Succession to Siemships in the Khasi Hills*, pp. 7-48.

37. *Ibid.*, Letter from F.J. Monahan to Secretary, Government of India, 22 April 1903.

38. *Bengal Administration Report 1871-72*, Statistical Returns, p. XII.

39. India Office Library and Records, London, Linlithgow Collection, MSS Eur 125/31, No. 51, R. Reid to Linlithgow, 11 November 1938. That Dakhor Singh, Syiem of Khyrim invited to attend the Delhi *Dorbar* of 1911 is evident of the position he held among the Khasi chiefs.

40. K. Cantlie, *Notes on Khasi Law*, reprinted, Shillong, n.d., p.144; *U Nongpynim*, October 1928, pp.1-19 and January 1929, pp. 15-21.

41. K. Cantlie, *Notes on Khasi Law*, p. 145.

42. ASR, Appointment and Political Proceedings, September 1931, No. 31.

43. David R. Syiemlieh, op. cit., pp 187-188. On the position of the *Syiems,* Sir Andrew Clow, the Governor of Assam referred to a note of the Legal Remembrancer Mr. Justice B. N. Rau who wrote in 1937: "I think it would be incorrect to say that the Siems have no sovereignty whatever: if that were so, their territory would be part of British India. In fact their judicial powers, such as they are, are in exercise of sovereign rights: the fact that those powers are limited merely means that their sovereign rights extended up to a certain point, after which the sovereign rights of the British Government came into operation-in other words it is a case of division of sovereignty". Sir Andrew Clow, 'The Future Government of the Assam Tribal People', in David R Syiemlieh (ed.), *On the Edge of Empire: Four British Plans for North East India,* Sage Publications, Delhi, p.212.

44. K. Cantlie, *Notes on Khasi Law*, p. 183.

45. India Office Library and Records, London, Linlithgow Collection, MSS Eur 125/31, No. 51, Reid to Linlithgow, 11 November 1938.

46. *Ibid.*

47. For details read David R. Syiemlieh, *British Administration in Meghalaya Policy and Pattern*, pp.195-205; Helen Giri, *The Khasis Under British Rule (1824-1947)*, Regency Publications, New Delhi,1998,pp.224-262.

8

The Federation of Khasi States: Three Phases of its History

Introduction

Political awareness emerged among the Khasi- Jaintia in the wake of the implementation of the Government of India Act 1919. Hitherto there had been little interest in modern politics other than the relations between the imperial government and the several Khasi states. The Jaintias had earlier established the Jaintia National Union in 1900. It was in its early years a socio-cultural group with little interest in politics. It did however play a role in recommending candidates for the 1937 and 1946 elections to the Assam State Legislature. Many years later it would become the Jaintia Durbar under which nomenclature its representatives would stand for election in the 1951 elections. More significant as a political body was the Khasi National Dorbar established in 1923. Its precursor was the Khasi- Jaintia United Guild. The Khasi National Dorbar included in its membership the Khasi *Syiems* and other traditional heads of the *himas*. These leaders joined with their fellow Khasis in the British areas in the district to discuss, adopt resolutions and become a pressure group in matters concerning their common situation. The first President of the Dorbar was *Syiem* Olim Singh of Khyrim. The founding Secretary was J. J. M. Nichols-Roy.[1]

While this development was under way in the Khasi Hills, progress had been made to give the Indian princes a political role. The Governor General, Lord Canning had referred to those states that had not joined in the uprising of 1857 as "breakwaters of the storm". He recorded that many Indian states if not neutral in the uprising, had given military support to the British. This situation so soon after many states hade been incorporated in the expanding British dominion in India was, if not a moral boost to the imperial power, a strong military support to suppress the other forces that had taken up arms against the government. Post 'Mutiny' British policy however, could not make use of the states as bulwarks, nor were they a social buttress as it was expected of the rulers. Governor-Generals Mayo, Lytton and Curzon at different times suggested the princes be given a forum to meet and discuss common issues.This ranged from the Council of Princes later suggested by Lord Minto to Morley, the Secretary of State for India to Minto's Council of Notables. When these suggestions could not take shape they reverted to the earlier suggestion to bring the princes together in a Council of Princes. What really set the pace of increased unequal partnership between the princes in general and the imperial power in India was the Princes response to the Secretary of State's August 1917 declaration in Parliament. The speech had reference to "India" which the Indian princes believed included them in the changes that were envisaged for India. The Secretary of State's visit to India in 1918 and the product of that visit and the report he and the Viceroy together brought worked out, encouraged in no small measure the inauguration of the Chamber of Princes on 8 February 1921.[2]

The early years of the Chamber of Princes did not include the participation of the larger Indian states in its deliberations. They were comfortable enough in their relations with the imperial power not to have an active role in the new body. The medium and smaller states needed an organisation of this nature and they were involved in its activities from the start. The participation of the Khasi states in the Chamber of Princes has not been researched. There are several reasons, which would explain this. The Chamber of Princes was useful to and dominated largely by the states of western and central India and the Punjab. The participation from the eastern states was more from Mayurbhanj, Cooch Behar, Manipur and Tripura, states that had the resources and size in their population and location to be heard in that body. The Khasi states were too small to have had individual participation. They are not known to have even had joint involvement in the Chamber of Princes.

Federation of Khasi States

The representatives of the Khasi states were on the other hand actively involved in the Khasi National Dorbar. The Dorbar at various sessions discussed issues relating to the relations between the states and the government and codification of land and inheritance laws in Khasi states.[3] Political activity in the hills was heightened from 1928 following the announcement in November and December 1927 that the Indian Statutory Commission and the Indian States Committee would review and make recommendations for changes in the administration of British India and the Indian states. Initially reluctant to submit memoranda to the Indian Statutory Commission as it was not the appropriate body to receive memoranda from the states, the Khasi chiefs together with other leaders welcomed Sir John Simon and other members of the Commission when they visited Shillong on 2 January 1929.[4] In preparation for the visit the Khasi National Dorbar had in May 1928 prepared and submitted a memorandum to the Indian Statutory Commission. The memorandum raised a number of issues relating in part to the Khasi States, the issue to their chiefs of *sanads* and the control by Government of wastelands in the states. The petition also suggested the establishment of a "central dorbar" as a federation of all the states for which was also submitted a draft constitution of the proposed dorbar.[5]Another memorandum was submitted to the Commission in January 1929. This made an appeal that the Khasis should be removed from the "backward category" of hill areas.[6]

The concept of the Federation of Khasi States first emerged in a meeting of the Khasi National Dorbar on 2 May 1929.S. G. Nalle initiated the discussion on the establishment and purpose of a federation[7].It was intended to unite all the 25 Khasi *himas* consisting of *Syiemships, Lyngdohships, Sirdarships* and the *Wahadadarship* under one organisation. The Constitution provided the election of office bearers to manage the affairs of the federation, the levying of taxes, and making of laws subject to the assent of the Governor of Assam. A decision was taken to print and circulate pamphlets to explain to the people the need to have a federation of Khasi states.

Three years passed before the federation took shape. Advantage was taken by the Khasi chiefs on the visit of the Governor-General, Lord Willingdon to Shillong in October 1933. By then the Indian States in general were again concerned about their political future. After the Simon Commission Report proposed an Indian federation of British India and the "Indian Native States", political activism was given fresh vigour. Several Indian states showed interest in the proposed Federation. The Khasi States in their memorandum to the Viceroy showed their interest in joining the Federation for which they looked forward to direct relations with the Viceroy through a Political Agent who would work exclusively for the States. In his reply to the memorandum of the Khasi states Lord Willingdon made mention of the "Republics" and their ability to maintain their freedom. In his address to the Khasi chiefs he said;[8]

> For some time now, you have been considering the feasibility of closer association amongst yourselves with a view to constitute a Federation of the Khasi States. I would commend this idea of your most earnest attention and this is obviously the first and most useful step which should pave the way towards your entry into the greater federation.

The visiting dignitary rejected the idea for placing the Khasi States in direct political relations with the Viceroy.

It is of some significance that the Federation of Khasi States, established in 1934 should have taken shape after the encouragement given to the chiefs by the Viceroy.[9] They needed a stimulant. That came with the Viceroy's reply to their memorandum. The situation within which it became possible to have the Federation should be noted. The chiefs expected much and wanted to play their part in the future developments. The aims and object of the federation were to discuss and consider political questions, to represent their "legitimate desires", to put forward their claims for a higher status in relation with the imperial power, to have greater judicial power and to show on all occasions their loyalty and allegiance to the British Crown.[10]

Despite the tremendous interest the leaders of the Khasi states and the political leaders in the British portions of the Khasi Hills showed in the establishment of the federation, it soon became inactive. The reasons why the Federation of Khasi States failed to continue as an organisation of the several *himas* were basically two. First, soon after its establishment, the Government of India Act 1935 was passed and brought into operation. The proposed Indian Federation provided in the Act, however did not become operative for numerous factors. This affected the hopes and aspiration of all the Indian states, the Khasi *himas* included. In substantial part this was the reason why the Federation of Khasi States went into limbo till the mid 1940s when it would be revived. Composed of small and financially weak states, the federation suffered from numerous organisational concerns, not the least being that the states had very few leaders other than the *Syiems* of Khyrim and Mylliem. And this was despite the discovery in 1933 that the Calcutta High Court had ruled in 1884 that the 25 Khasi chiefs were "Ruling Chiefs."[11]

Revived Federation

By 1945 it had become clear that the British would leave India after the War. The Indian States were once again activated when the Cabinet Mission submitted its Memorandum of State's Treaties and Paramountcy of 22 May 1946.An important constitutional and political issue for the British Government and the Interim Government which would assume office in September of that year, was the determination of the future of the Indian states after the lapse of British paramountcy. The Memorandum stated that with the transfer of power the British Crown would cease to exercise paramountcy and that the rights of the Indian states would return to them. The void that would arise from the lapse of political arrangements between the states and the Crown could be filled in either with the states entering into federal relationships with the succeeding Government of Governments or enter into political arrangements with or without them. The States were free to associate with one or the other of the Dominions of India and Pakistan, to federate among themselves or to stand alone.[12]

It is in this background that the Federation of Khasi States was revived on 22 August 1946. The establishment of the Federation came at the apogee of the British Empire in India. Its revival was ushered in the collapse of that structure and its fallout. As before in the 1920s-1930s, the Khasi leaders were divided on the question of the future of the States. Some wanted amalgamation with the British villages and that the District of Khasi and Jaintia Hills should become part of Assam. Others preferred that the Khasi and Jaintia Hills be united but that the hills should not be part of Assam. There were still other who wanted no merger and integration with either Assam or India. The details of these developments from the signing of the Standstill Agreement in July 1947; the Instrument of Accession in December that same year and in the months following; the complex issues arising from the signing of these documents and the question of the merger of the Khasi States into India have been extensively researched and published. [13]

The Adviser to the Governor of Assam for Tribal Areas and States was seized with the situation arising out of all these developments. He wanted immediate negotiations between the representatives of the several states and the Provincial Government. He did not want piecemeal arrangements with each State, as it was undesirable. To make matters easier in negotiations it was suggested that the Federation of Khasi States should be recognised as the body representing the States. Some of the concerns of the Adviser in the matter of the Khasi states were those relating to the position of the

Shillong urban area; law and order; the operation of laws; the continuation of trade, regulation of forests; supplies; transport; mining; and the possible return of Cantonment land to the *Syiem* of Mylliem.[14]

These and other matters were discussed between the Government and the Federation of Khasi States. The Federation had an office in Riatsamthiah, Shillong. Inaugurated on 15 June 1947 by the first Indian Governor of Assam, Sir Akbar Hydari, with a flag of twenty-five stars (representing the 25 Khasi states) as a symbol of its authority, the Federation was to operate with a legislature, a judiciary and an executive. There was very little operation of administration under these provisions, as the Federation had taken up larger issues relating to the future of the Khasi states in an independent India. The "transition period" as Homiwell Lyngdoh called it covered the period from 15 August to the adoption of the Constitution of India on 26 January 1950.[15] Within this period the Federation of Khasi States operated under the provisions of agreements reached between them individually and severally and the Government.

Third Federation

Surprisingly there seemed to have been no adverse reaction from the Khasis in general and their chiefs in particular to their district being integration into India and tagged on to Assam. It was a silent acquiescence. This acceptance of their position continued till very recent times.

The second Federation of the Khasi States was revived before the transfer of power from the British to independent India. It too became moribund and for a much longer period than its precursor. The third Federation of Khasi states was revived in 2000 and under very different circumstances. The background for the revival of the Federation lay in part to the events shortly before and after the transfer of power, of going back to history in a sense and reviewing the process by which integration of the Khasi States had been incorporated into India. In recent times there has been the more active participation of students as young ombudsmen in the state of Meghalaya. Alongside this development is the assertion by the traditional chiefs, the *Syiems, Sirdars, Lyngdohs* and *Wahadadars*, questioning their position in their relations with the Khasi Hills District Council, particularly over a legislation of 1959 which reduced them to headmen of their respective *himas* and functionaries of the District Council. There is a call for providing the traditional leaders constitutional recognition and funds.

Notes and References

*David R. Syiemlieh,"The Federation of Khasi States: A Study of Three Phases of its History," *Proceedings of the North-East India History Association*,Twenty-fourth session, Guwahati, Shillong, 2004, pp.380-388.

1. For a general reading on the political developments in the Khasi- Jaintia Hills see, Amalendu Guha, *Planter Raj to Swaraj: Freedom Struggle and Electoral Politics in Assam 1826-1947*, Peoples Publishing House, New Delhi, 1977, reprinted,New Delhi, 198 and S. K. Chaube, *Hill Politics In Northeast India*, Calcutta, 1973, reprinted Orient BlackSwan, N. Delhi, 2001.

2. S. R. Ashton, *British Policy Towards the Indian States,* Curzon Press, London and

Dublin, 1982. See Chapters 1 and 2.

3. Evalyn Ivyleena Nongkynrih, "History of the Growth and Development of Political Awareness in the Khasi- Jaintia Hills (1900-1950)", M. Phil dissertation, NEHU, 1993, pp.52-56.

4. *Ibid.* p. 57; *David R. Syiemlieh, British Administration in Meghalaya: Policy and Pattern*, Heritage Publications, New Delhi, 1989, p. 178.

5. E. I. Nongkynrih, op. cit., pp. 57-58; David R. Syiemlieh, op. cit., pp. 178-179; *U Nongpynim*, December 1928, pp1-6.

6. *U Nongpynim*, December 1928, pp. 18-19.

7. Helen Giri, *The Khasis Under British Rule 1824-1947*, Akashi Book Stall, Shillong, 1980, p. 133.

8. Memorandum submitted to the Viceroy Lord Willingdon on his visit to Shillong, 3 October 1933.

9. David R. Syiemlieh, op. cit., pp. 177-178.

10. *Ibid.*; Keith Cantlie, *Notes on Khasi Law*, Shillong, 1934, reprinted Shillong, (n.d.).Keith Cantlie, the Deputy Commissioner of the Khasi and Jaintia Hills District and Political Agent in relation with the Khasi States made suggestions for the proper working of the Federation.

11. India Office Library and Records, London, Linlithgow Collection, MSS Eur F 125/31, No. 51, Reid to Linlithgow, 11 November 1938.

12. Nicholas Mansergh (ed.), *The Transfer of Power*, HMSO, vol. VII,No. 262, pp. 522-524.

13. Helen Giri, op cit.; E.I. Nongkynrih, op. cit., David R. Syiemlieh, op. cit.

14. Assam Secretariat Records, Assam Governor's Secretariat, File A 1701/47, Notes on Khasi States.

15. Homiwell Lyngdoh, *Ki Syiem Khasi Bad Synteng*, Ri Khasi Press, Shillong 1964, p.v.

9

The Integration of the Khasi States into the Indian Union

Introduction

Much has been written on the integration of Indian states into the Indian Union following Independence. However, very little attention is drawn to North-East India where there were a number of states-Manipur, Tripura and the Khasi states. Some of these states abutted on East Pakistan. Source material on the process of integration of these states is abundant, primarily in archival material that unfortunately has not been opened for study. Researchers have thus had to depend on whatever official correspondence is available either in original or in edited form, on memoirs, newspapers, some pamphlets, and secondary literature. This essay relates to the Khasi states alone. No doubt some work has been done in this field. It is hoped that this essay shall make a contribution in giving a somewhat more complete picture of the integration process of the Khasi states as also to indicate whatever material is presently available to understand the problem.

The Transfer of Power

The transfer of power from Britain to the Dominions of India and Pakistan was first discussed in the Cabinet Mission Plan of 16 May, 1946 in which the Cabinet Mission and the Viceroy, in consultation with the British Government issued a statement embodying their suggestions and recommendations towards a solution of the Indian political question. The most important constitutional issue then was to determine the position and future of the 550 odd Indian states.

Referring to these states the Cabinet Mission said that with the attainment of independence by British India the relationship which had existed between the states and the British Crown would no longer be possible, though it was expected of the states to co-operate with the new governments in building up a new constitutional structure.[1] The position of the states was further elucidated by the Cabinet Mission in its Memorandum on States' Treaties and Paramountcy of 22 May, 1946. The Memorandum stated that with the transfer of power His Majesty's Government would cease to exercise paramountcy. This meant that the rights of the states in relationship with the Crown would no longer exist and that all rights surrendered by the states to the paramount power would return to the states. The void that would arise from the lapse of political arrangements between the states and the Crown was to be filled in either with the states entering into federal relationship with the succeeding governments or enter into political arrangements with or without them. States were, therefore, free to associate with one or the other Dominions, to federate among themselves or to stand alone. The British Government emphatically stated that it would not put the slightest pressure or influence in deciding which Dominion the states should accede to.[2] Realizing that the states would find it difficult to exist independently, the Secretary of State for India underscored the importance of states to find their appropriate place within one or the other of the two new Dominions.[3]

On 15 August, 1947 British rule in India ended and erstwhile British India was partitioned. Earlier an interim Government was sworn in on 2 September, 1946.Its function was to fill in the time gap pending the framing of a new constitution for India. The responsibility of negotiating with the states to accede into India was entrusted to the States Department of this Government. To remove all possible fears and suspicions in the minds of the Indian rulers, Sardar Valabhbhai Patel who headed this Department issued a statement underlying the paramount necessity of maintaining the unity of the country by the states joining the Indian Union for defense, foreign affairs and, communications. He admitted "it is an accident that some live in the States and some in British India". Although with the transfer of power, paramountcy would lapse, he urged that it was in the interest of India and the Indian states that, the working of the treaties and agreements entered by the states with the British Government should continue to operate until new agreements were made.[4]

The Khasi States

In their relations with the Khasi states the British recognized twenty-five states categorised as semi-independent and dependent states. These states had from the third decade of the nineteenth century entered into relations with the East India Company and the Government of British India through *sanads*, engagements and *parwanas* which clearly laid down the principles of relations between the two parties. Juxtaposed to or interspersed with these states were thirty-five British villages in the Khasi Hills which formed part of British India. The Jaintia Hills, formerly part of the Jaintia *raj* was annexed into British India in 1835. The emergence of political activity in these hills in the early part of the last century affected the rulers of the states. In early 1934 was formed the Federation of Khasi States.[5] The Federation was revived in 1946 when the state became concerned about their future. It was with this Federation and the individual Khasi chiefs that the Indian Dominion had to negotiate for their integration into the Indian Union.

In early April, 1945 it was reported that the tribal people were beginning to take a more vocal interest in their own future. A meeting in Shillong which contained most of the more prominent men opposed emphatically their inclusion in either Pakistan or India.[6] But the future of these hills was not to be decided by this body but by the Khasi chiefs. Sometime in July, 1947 an agreement was reached between the states and Sir Akbar Hydari, the Governor of Assam on the three terms that Patel had asked the states to accept.[7] On 9 August the Khasi states signed the Standstill Agreement. They agreed that with effect from 15 August, 1947 all existing arrangements between the Province of Assam and the Indian Dominion on the one hand and the Khasi states on the other should continue to be in force for a period of two years or until new or modified arrangements should be arrived at between the authorities concerned. The agreement was subject to certain exceptions which gave the federated states judicial, administrative, legislative and revenue powers. It was also agreed that all British villages in the district which decided to rejoin states of which they formerly formed a part should be allowed to do so.[8]

The Government of India found a problem when it came to getting the Khasi states to sign the Instrument of Accession. On 2 December that year Governor Akbar Hydari informed the Khasi chiefs that he had brought with him from Delhi the Instrument of Accession and that they should sign it. It was accordingly agreed that all the twenty-five chiefs should assemble at the Governor's residence on 15 December and individually sign the Instrument. Twenty chiefs signed the Instrument that day, among the remaining five states the chiefs of three were ill and would sign at home, while two refused to sign, it being assumed that summons had not reached them.[9] Later that day Hydari reported to Patel:[10]

> That various underhand forces had been at work between 2 December and 15 December is shown by the fact that this morning's proceedings seemed likely to break, for, three of the principal Syiem; i.e. those of Mylliem, of Khyrim and of Cherra refused to sign and wanted more time "to consult their people." I made them realise what the consequences of not signing would be, and after nearly an hour's confabulation among themselves they signed. The rest was easy.

Generally the Khasi states had no desire to join Pakistan. The *Syiem* of Cherra did 'flirt' with the local authorities in Sylhet before signing the Instrument of Accession but was warned by Hydari against playing that game. The *Syiem* was attached to Pakistan for the simple reason that some part of his personal land lay in Sylhet. Hydari had exerted his authority during these negotiations by intimating the chiefs that the fact of mere accession was not a guarantee of a particular person continuing as a chief and that if there was substantial amount of feeling in a particular state that its chief was not doing his duty, he would have an enquiry conducted by the Deputy Commissioner. If it was found that allegations against a chief were true fresh elections would be ordered. This undertaking by the Governor reconciled the people at large to the signing of the Instrument of Accession. This was not taken well by the chiefs who found their tenure thereby insecure. The Federation considered it as a diminution of its influence.[11]

Nobosohphoh and Nongspung states signed the Instrument on 11 January, 1948 followed by Mawlong on 10 March.[12] There remained Rambrai and Nongstoin which procrastinated. Hydari then sent G.P.Jarman, the Deputy Commissioner/Dominion Agent and his Assistant, R.T. Rymbai to these states with instructions that failure to comply to signing the Instrument would be followed by pressure of various kinds and in the last resort to deposition.[13] At one time, it looked as if Jarman might encounter armed opposition from Nongstoin and so a platoon of the Assam Rifles was sent into the state "whose presence and Jarman's tact did the trick."[14] The *Syiem* signed the Instrument of Accession on 19 March, 1948. Rambrai had signed two days earlier.[15] Governor Hydari who felt that the policy should be one of conciliation and patient adjustment of difficulties is said to have told R. T. Rymbai before leaving for Nongstoin, "Let Junagadh not be repeated.[16] Behind a tough exterior Hydari had a concern that there should be no violence in the integration process.

More material is now available about how Nongstoin acceded into India. R.W. Selby, the British High Commissioner to India had come across a curious reference in *The Sunday Statesman* of 28 March, 1948 to an alleged appeal by the *Syiem* to the Security Council of the United Nations Organisation against the unlawful aggression of the Indian Government into his state. The *Syiem* was also understood to have sent a note to Jawaharlal Nehru requesting the withdrawal of Indian troops "in order to avoid further complications". Wickliffe, the *Syiem's* nephew and likely successor to the *Syiemship* who disclosed this was then preparing to leave for Lake Success to take up the matter with the United Nations Organization.[17] Selby's enquiries made in official quarters confirmed that the *Syiem* had in fact sent a letter to Nehru. There was however no reference in the *Syiem's* letter of appeal to the U.N.O.[18] Further enquires revealed that on the day the *Syiem* signed the Instrument of Accession certain parties, official reports say, were sending out false telegraphic reports to the effect that the Government had sent military forces into the state and that the chief had appealed to the U.N.O.[19] That Nongsoin and Rambrai were pressurised into acceding into India there is no doubt. The appeal to the World body must still be verified. Wickliffe does not appear to have gone to the U.N.O. It is believed that he remained in East Pakistan.

Issues with Merger

The Khasi states had acceded into India but refused to merge on the ground that the chiefs were elected heads of their respective states. Their refusal caused Sardar Valabhhai Patel to visit Shillong on 1-2 January, 1948. His meeting with the chiefs ended in a stalemate over the merger issue, for the Khasis said that only a duly constituted *dorbar* of the states could decide on such a decision.[20] Accordingly rules were drawn up by the Dominion Agent for the nomination and election of members of the Khasi States Constitution Making Durbar.[21] This took almost sixteen months. The Durbar was inaugurated on 29 April, 1949.

While the Khasi States Constitution Making Durbar had just been convened the Indian Constituent Assembly was preparing the final draft of the Constitution. J.J.M. Nichols-Roy who was a member of both the Assembly and the Durbar urged the latter to accept the broad framework of the Sixth Schedule of the Constitution. The Schedule was the product of the North East Frontier (Assam) Tribal and excluded Areas Sub-Committee headed by Gopinath Bordoloi, the Assam Premier. Its report submitted to the Constituent Assembly on 28 July, 1947 had pointed out that the Khasi states had comparatively little revenue or authority and seemed to depend for a good deal of support on the Political Officer in their relations with their people. It believed that there was a strong desire among people of the states to "federate" with the people of non-state villages. It was also noted that some of the *Syiems* favoured amalgamation but their idea of the Federation differed from that of the people in that the chiefs sought greater power for themselves than what the people were prepared to concede to them.[22] By then factionalism had raised its head in the Khasi Hills with two political bodies vying with each other to voice the demands of the people. The Federation had as its agenda the interest of the states. The more popular body was Reverend J. J. M. Nichols-Roy's Khasi-Jaintia Federated State National Conference. By then Nichols-Roy was much disliked by the chiefs for the official stand he was taking.

The question of the future administrative arrangements for the Khasi and Jaintia Hills was a matter of much concern. On 21 July, 1949 Dr. Homiwell Lyngdoh, the Chairman of the Durbar read Nichols Roy's resolution which suggested the formation of an autonomous unit of the Khasi and Jaintia Hills within Assam province. This was followed by the *Syiem* of Jirang's amendment demanding one united administration for the two hills outside Assam, provision for which was possible under the terms of the Instrument of Accession and the draft Constitution of India. Though Nichols-Roy and his supporters had a majority of members in the Durbar, the vote over the resolution and its amendment went with a 40-46 victory for the chiefs. At this Nichols-Roy and his 39 supporters over the debate walked out of the proceedings of the Durbar. The remaining members then elected a sixteen member Negotiating Committee which sent a resolution to the Drafting Committee of the Constituent Assembly for a reconsideration of the future status of the Khasi states and non-state villages.[23]

The twenty-five Khasi states were too small, even collectively to get representation in the Constituent Assembly. The eminent anthropologist, G.S. Guha was made the representative of the Khasi states, Tripura and Manipur. The Khasi chiefs suffered two disadvantages. Guha does not appear to have said anything on their behalf. Nichols-Roy did all he could to undo the defeat he had suffered in the Durbar. On 7 September a resolution was adopted at the Constituent Assembly creating the United Khasi-Jaintia Hills District comprising the territories which before the commencement of the Constitution were known as the Khasi states and non-states areas.[24] There was still no mention in this third reading of the draft Constitution of whether the district would form part of Assam province. Nichols-Roy was particularly happy that the Khasi states had been incorporated in the Sixth Schedule for it would enable the same people (apart from being a personal triumph) to have one administration for the two categories of areas.[25]

Mohammad Saadulla, earlier Assam's premiers pointed out an anomaly over what had been accepted. "Sir", he addressed the Chairman of the Constituent Assembly:[26]

The Khasi Hills have been relegated to the Sixth Schedule for which Rev. Nichols-Roy is very thankful, but there is a constitutional anomaly. Although the Constituent Assembly is not to find a remedy for that, yet I must sound a note of warning that this small district of Khasi Hills embrace 25 Native States most of which had treaty rights with the sovereign power in Delhi. They were asked to join the Indian Dominion in 1947. Instruments of Accession accompanied by an Agreement were executed by these chiefs and they were accepted by the Central Government. But even though this area has been included in the Sixth Schedule, up till now no agreement or settlement has been arrived at between the Constituent Assembly of the Federation of the Khasi States and the Assam Government or the Government of India.

Saadulla added that Olim Singh, President of the Federation of Khasi States had led a delegation early in November to Delhi to press their grievances before the States Ministry and the Drafting Committee. But he said: "they are late in the day and nothing can be done at the third reading."[27] The Draft Constitution was adopted on 26 November 1949 and the Assembly was adjourned till 26 January, 1950.

A year earlier a suit was filed in the Federal Court of India by Sati Raja, *Syiem* of Mylliem, against the Dominion of India and the Assam Government. Carefully worded the case reviewed the developments between Mylliem and the Indian Dominion since the Standstill Agreement and the arbitrary manner of the two Governments in continuing to exercise the rights, privileges and jurisdictions that the former Government had exercised. Among others, the *Syiem* wanted a declaration that his state had recovered or was entitled to recover sovereign rights, power, functions and jurisdiction over his state.[28] This must have been quite an embarrassment for the Governments of India and Assam. Sri Prakasa, the Governor of Assam met Sati Raja and others on 31 December, 1949 and was able to make the *Syiem* withdraw his case.[29]

Something of the irregular manner of administering the Khasi Hills that the Syiem of Mylliem sought clarification require to be explained. On 15 August, 1947 the Governor-General issued a Provisional Constitutional Order abolishing all references to "tribal areas" and the distinction between "India" and "British India". This was followed on 27 August by the Extra-Provincial Jurisdiction Ordinance re-establishing retrospectively the severed links which resulted from the first order. Two notifications were issued under this Ordinance. The first promulgated the Assam Tribal Area Order, 1947, confirming and giving effect to every instrument, that is, every notification, order, bye-law, rule, regulation or directive made or issued under Section 313 of the Government of India Act, 1935. The second notification authorized the Assam Governor to continue to discharge his former functions in or in relation to the tribal areas in Assam as the Agent of the Governor-General.[30] Under provisions of the Extra-Provincial Jurisdiction Order, 1947, a special notification, the Khasi States Federation (Administration of Justice) Order, 1948 was made applicable from 1 July. This notification defined the civil and criminal powers allocated to the Federation of Khasi States and the Khasi States under the supervision of the Assam High Court.[31] A Khasi Federation Court and Executive began exercising the judicial and executive functions formerly vested in the Deputy Commissioner, who however remained in that capacity and as Dominion Agent. This arrangement continued till the end of 1949. One may surmise that very little control was actually transferred to the chiefs whose future continued to be undefined.

More surprise was in store for the Federation and its chiefs. One day before the Constitution of India was adopted the Governor of Assam passed an order cancelling the Khasi States Federation (Administration of Justice) Order, 1948 and its Supplement of the same year. The Khasi States (Administration of Justice) Order, 1950 which came into force on 25 January, 1950 entrusted civil and criminal justice to the Deputy Commissioner of Khasi and Jaintia Hills District, his Assistant and the Courts of the *Syiems*, *Sirdars*, *Lyngdohs* and *Wahadadars* in similar manner to the pattern that existed during British administration. That same day another notification was issued changing the designations of the Dominion Agent, Additional Dominion Agent and the Assistant to the Dominion Agent and the Court of the Khasi States Federation as referring respectively to the Deputy Commissioner, Additional Deputy Commissioner, Assistant to the Deputy Commissioner and the Court of the Deputy Commissioner, Khasi and Jaintia Hills district.[32] The integration process was complete.

Apparently this was done to conform to the Constitution for Part A of the First Schedule of the Constitution read that the territory of the Assam "shall comprise territories which immediately before the commencement of this Constitution were comprised in the Province of Assam, the Khasi States and the Assam Tribal Areas." Thus the Khasi states became part of Assam without any agreement of merger and disregarding the provisions of the Standstill Agreement. It was in the Constitution of India, the drafting of which the chiefs played no part that the integration of the Khasi states into India was made complete. The final process may have been done arbitrarily, but this was only possible because of the divergent views among the Khasis, their indecision and delay to merge, all of which was taken advantage by those who held the ropes in Delhi and Shillong.

Notes and References

*David R. Syiemlieh, 'The Integration of the Khasi States into the Indian Union,' Sajal Nag *et al.* (ed.), *Making of the Indian Union Merger of Princely States and Excluded Areas*, Akansha Publishing House, New Delhi,2007.

1.   N.Mansergh (ed.), *The Transfer of Power 1942-1947*, vol. vii, HMSO, New Delhi, 1978, No. 303, p. 586.

2.   *Ibid.*, No. 262, pp. 522-524.

3.   B.S. Rao (ed.), *The Framing of India's Constitution*, vol. i, New Delhi, 1968, p. 534.

4.   S.L. Poplai (ed.), *India 1947-1950 (Select Documents on Asian Affairs)*,vol. i, Bombay, 1959, No. 39, p. 170.

5.   K.Cantlie, *Notes on Khasi Law*, Reprint, Shillong, n.d., pp. 176-186; H. Bareh, *The History and Culture of the Khasi People*, Calcutta, 1967, p. 235.

6.   N. Mansergh, op.cit, vol. v; London, 1974, No. 397, p. 912.

7.   Durga Das (ed.), *Sardar Patel's Correspondence*, vol. v. Ahmedabad, 1973, No. 43, pp. 42-43.

8. L.L.D. Basan, *The Khasi States under the Indian Union*, Shillong, 1948, pp.1-3.

9. Durga Das, op. cit., pp. 42-44. These five states were Nobosohphoh, Nongspung, Mawlong, Rambrai and Nongstoin.

10. *Ibid.*, p, 43.

11. Durga Das, op. cit.,vol. vi, no. 74,Ahmedabad, 1973, pp. 101 103.

12. S. R. Poplai, op. cit., No. 55, p. 236; *White Paper on Indian States*, New Delhi, 1950, p. 216.

13. Durga Das, op.cit., vol vi, No. 74, p. 103.

14. *Ibid*, No. 76, p. 105; Durga Das, op. cit. vol. vi, No. 76. p. 105.

15. *White Paper on Indian States*, p. 216; S. R. Poplai, op.cit., p. 236.

16. Durga Das, op.cit., vol. vi, No. 74, p. 103; Interview with R.T. Rymbai, 8 July 1979.

17. India Office Records, London, L/P&J/7-10635,R.W.Selby to H.A.F. Rumbold, 13 April,1948; *The Statesman*, 28 March, 1948.

18. *Ibid.*

19. *Ibid.*, Selby to Rumbold, 10 May, 1948; V.P.Menon, Secretary of the States Department gave the impression to Selby that he considered the whole affair as having no importance at all. The Governor-General accepted these Instruments of Accession on 17 August, 1948. India Office Library and Records, London, L/P&J/7-10635, W. R. Selby to H. A. F. Rumbold, 10 May, 1948; White Paper on Indian States, p.216.

20. H. Bareh, op.cit., p. 241.

21. L.L.D. Basan, op. cit., pp. 29-35.

22. B. S. Rao, op. cit., vol.iii, New Delhi, 1969, p. 688.

23. L. G. Shullai, *Ki Hima Khasi*, Shillong, 1975, pp.11-13.

24. *Constituent Assembly Debates*, vol.ix, reprint, New Delhi, 1967, p. 1008.

25. *Ibid.*, p. 1009.

26. *Ibid.*, vol. xi, reprint, New Delhi, 1967, p. 737.

27. *Ibid.*

28. Case No. V of 1949 in the Federal Court of India, New Delhi.

29. L.G. Shullai, *Ka Ri Shong Pdeng Pyrthei*, Shillong, 1978, p.15.

30. S.C. Chaube, *Hill Politics in Northeast India*, third edition, Orient BlackSwan, New Delhi, 2012 pp. 85-86.

31. L.L. D. Basan, op. cit., pp. 12-28. On 4 October 1948, the Ministry of States issued a Supplementary Notification to the Khasi States Federation (Administration of Justice) Order, 1948.

32. L. G. Shullai, op.cit., pp. 27-36.

10

Call of Freedom from the Hills: Tirot Sing and his Significance in the Freedom Struggle

Introduction

History teachers in the North East region find it hard to explain to the more inquisitive student why, if they have to teach the struggle of Tipu Sultan against British colonialism and Bipan Chandra Pal and his vision of a free India and many more personalities, there is no requirement for students to read on regional personalities such as Tirot Sing, the Khasi chief who resisted British imperialism; Maniram Dewan of Assam who suffered trial and was sentenced to death, and accounts of other resistance movements against the colonial state in the region. The impression created reading books such as Bipan Chandra's *India's Struggle for Independence* and

R. Suntharalingam's *Indian Nationalism: An Historical Analysis*, to mention only two references, is that the region did not have anything to do with the Indian freedom movement. Not even the widely acclaimed and readable *Freedom Struggle* has reference to any political participation beyond Bengal. A corrective would have to wait for Sumit Sarkar's prize winning *Modern India 1885-1947* for mentioning significant strands of the history of the region within the broader framework of his study. Amalendu Guha corrected this lapse in his Presidential Address in the Modern Indian History section of the Indian History Congress in 1983 when he wove aspects of the history of the North East into his discussion on Indian nationalism in general. Some year later J.B. Bhattacharjee in his address to the members of the Indian History Congress at Gorakhpur, 1989 dealt at length with 'World War II and India, the Bose-Gandhi controversy and Transfer of Power' with focus on the region.

On the other hand the histories of the North East is largely insular, with attention on the numerous tribe, their early states. These histories have not been integrated into the general histories of the country. H.K. Barpujari the first President of NEIHA cautioned members of the Association in writing the history of the region:[1]

> There has been a general tendency in our local studies to magnify regional achievements, to exalt local heroes and to glorify local languages, customs and usages out of proportion to their intrinsic value.

He urged that regional study should be "complimentary not competitive," and that it should be taken as part of Indian history as a whole. He closed that session of his address by saying that a comprehensive history of India would be possible only on the bedrock of regional history.

Reinterpretation

One of the wonderful things of the subject matter of history is that it continues to be rewritten, more so when historical material becomes available to rewrite the past. Events and issues in the present also influence a fresh look and reinterpretation of the past. With this background the present interest of the Khasis regarding Tirot Sing, the *Syiem* of *hima* Nongkhlaw, one of the twenty- five Khasi states becomes significant.

So much has been said and written on Tirot Sing, *Syiem* of *hima* Nongkhlaw and of his struggle against the British colonizers, that his life and death story has become a saga. There are two broad perceptions on the man. The British in almost all their official and non-official references to him portray a particular perception, prejudiced by their belief that they had the right to rule and their racist views on subject people. In sharp contrast is the other perception of the man-contemporary and local. While the perceptions of the British writers was grossly prejudiced, the present euphoria on the man and his mission while closer to the truth, has resulted in a number of myths being woven around Tirot Sing and his death in particular.

Tirot's involvement in the Nongkhlaw 'massacre' of 4 April 1829 has not been fully established. It was reported that he was connected with the massacre of two British officers and other natives. One of the first official accounts of the Nongkhlaw incident of 1829 and its aftermath-the Khasi struggle and resistance to British rule, was written by R.B. Pemberton and published in 1835. He wrote the event of early April of that year "as an act the most atrocious cruelty." He went on to describe the Khasis as savages and the "atrocious conspiracy, diabolical cruelty of these misguided and infuriated savages." Many years later Alexander Mackenzie, the official historian of the British interest in the North East could not but use the same descriptive words to narrate the background and events in these hills between 1829 and 1834. References to this event in the Khasis hills by missionaries William Robinson and Alexander Lish too were coloured by their nationality. The latter writing from Cherrapunji some years after the Khasis were subjugated, makes mention of the "inhuman acts committed under his rule." This refrain contained well into the early years of the last century. The *Gazetteer of Assam* described the advent of the struggle of the Khasis as "wanton outrage", of "treacherous and suspicious Khasis." So much for this version of the Anglo Khasi War.[2]

Little effort has been made to understand why a free people as the Khasis were forced by circumstances to take up arms against the British. There is no literature to establish the Khasi view of these events but we may deduce that their *Syiems* and people must have been concerned of the gradual penetration of the British into their hills in the early decades of the 19th century. The construction of a sanatorium at Cherrapunji and Nongkhlaw was well under way by 1829. Of more concern for the Khasis was the construction of the road by the British, connecting Sylhet in the southern plains with Kamrup to their north and the arrival of convicts from the plains to construct that road. By 1826, the northern foothills were annexed to the territories of the East India Company. The Sylhet plains where a number of Khasi *himas* (states) controlled land had already come under the Company rule in 1765. Moreover the Company official had cleverly entered into treaties with several of the Khasi *Syiems* in matters of defense, road connection and sanatoria. The Khasis might not have realized the implications of the treaties at the time of giving their consent. They might not even have understood what they were entering into, as not all were familiar with the Bengali language in which the early treaties were signed. The official explanation for the outbreak has been ascribed to the "false and foolish speech of a Bengallee chupprassee, who, in a dispute with the Cassyas .. threatened them with his master's vengeance, and had plainly told them that it entered into his master's plans to subject them to taxation, the same as the inhabitants of the plains."[3] Many Khasi *himas* participated in the confederacy to take up arms against the British. The resistance was a protracted one. For over three years the Khasis fought the *Firingi-* (*Phareng* in Khasi) with fortunes of the "war" and the despair of loss of human lives occurring to both the Khasis and the Company forces.

The second and local perception of *Syiem* Tirot Sing begins to emerge before the culmination of the Indian people's freedom struggle and from shortly thereafter. First published in 1938, Homiwell Lyngdoh's *Ki Syiem Khasi Bad Synteng* provides short histories of each of the Khasi *himas*. It also provides the genealogy of the Syiemships after the pioneering report of D. Herbert's *Succession to Siemships in the Khasi States*. Homiwell Lyngdoh mentions the 1829 "massacre" without comments as it would have not been appropriate for him to have done so considering his service as a medical doctor under British rule. He does mention however that after his surrender in January 1833, *Syiem* Tirot Sing was taken to Dacca where he passed away on 29 March 1834 after a stomach disorder. From where did Lyngdoh get his information, which as will be shown, was well off the mark on the date of demise? Apparently he had some clue that the *Syiem* passed away some months after arriving in Dacca. But why then was nothing more recorded in Khasi oral tradition on the last days of the man? Tradition also says that his son and kinsmen visited him before his death. What is of concern is that though the reason for his death is known and that he had visitors from the hills, there is no tradition of the cremation/burial of the *Syiem* or where his mortal remains were laid to rest. Should his son have visited him in Dacca was this son from a Khasi mother, for another tradition say Tirot Sing had married someone related to the *Nawab* of Dacca. The mortal remains of the *Syiems* of Nongkhlaw were by tradition placed in *mawshyieng* at Mawmluh near Sohra. It becomes apparent that if there are no tradition or official sources on the burial of the bones in this location, then in all likelihood this was not done. How true is it that Tirot Sing's mother, *Ka* Ksan saved the life of David Scott by warning him that the Khasis would attack the Nongkhlaw sanatorium?

On 16 December 1952, Jairamdas Doulatram, the Governor of Assam laid the foundation stone of the Tirot Sing Memorial at Mairang. On the occasion he spoke of the rare courage and dignity of the *Syiem* and hoped that his name would find its place in the history of India's independence. The Governor requested a "second memorial' in the form of a good biography. On 29 March 1954, the same Governor unveiled the monument on what was then believed to have been the 120th death anniversary of Tirot Sing. The background in which the people of Nongkhlaw erected the memorial was that of the independence of India. Behind them was the national movement and India's independence. In this background the role played by resistance leaders was highly relevant and carried strong emotional overtones.

The legend of Tirot Sing has since grown. It has found expression in many forms, in songs, poetry, drama, art and biographies. Tributes have been paid to the man in the songs of late Elkin Swer, T.T. Mukhim, Rana Kharkongor, Lis Syiemlieh, Skendrowell Syiemlieh and Chosterfield Khongwir to name a few. There are three artist impressions of what he could have looked like. Khasi literature has a number of dramas on the *Syiem*. In 1956 was published in English, V.G. Bareh's *U Tirot Sing* of which more will be said shortly. Two short dramas of the Rympei Theatrical Centre; R.G. Phankon's, *Ka Sgni Khadduh* and H.A.M. Nongrum's *Ka Kput Kylliang* formed the base for Reginald Nongkynrih's *Ka Bniat Namar Ka Bniat*, published in 1985 and first staged that same year. It is informed that a Bengali theatrical group from Calcutta staged a drama on Tirot Sing in Shillong and other centre in these hills in the late-1940s. The legend then is not only pushed back in time, it becomes the theme of artists from Bengal.

Jairamdas Doulatram's suggestion that an account be written took some years before it was acted upon and a number of biographies were published. I have in mind four studies in English. Hipshon Roy's sketch was intended to take the name of this Khasi leader beyond these hills. The foreword to the tract described the man as, "one of the gems among the patriots," and that he had become "a martyr by his death in prison." The author considered U Tirot Sing as, "one of the most heroic but little known figures in the History of India." Jerly Tariang has two books on the Khasi leader. *U Tirot Sing* was published in September 1982. A more detailed biography, it has added to our knowledge of Tirot Sing and his times. In 1990 he published *Tirot Sing*. Till date the most widely circulated and the most detailed account of the Anglo-Khasi war and the life and death of this Khasi leader is Hamlet Bareh's U Tirot Singh (1984). Published by the Ministry of Information and Broadcasting under the *'Builders of India'* series, the book has had wide publicity and has taken the history of the struggle of the Khasis well beyond their hills.

Last Days

After his surrender in January 1833, Tirot Sing was sent to Gauhati to stand trial in the Foujdary Court. He was initially to have been sent to Tenasserim in Burma just recently annexed to the Company's Indian territories. The order was revised that he be lodged in the Dacca jail. By late February 1833 Tirot Sing was in Dacca in an apartment in the common jail.

At a lecture in the North-Eastern Hill University in early-1988 I read a paper,[4] drawn largely from material collected from India Office Library and Records, London, which was afterward distributed, in which was first uncovered the date of death of the Khasi *Syiem*. Searching through the Bengal Judicial Consultations I located a letter from S. C. Scott, Officiating Magistrate of the Foujdary Adalat, Dacca informing the Officiating Commissioner of Circuit, Dacca of "the demise of the ex-Rajah Teeruth Sing, a state prisoner under my charge, which event took place yesterday at one p.m." Scott's letter was dated 18th July 1835 which makes 17th July the date of death of the Khasi chief. This lecture was followed periodically with articles in newspapers and journals on the last days of the man. The intention was to clear the controversy over the date of death and to inform the public that the *Syiem* was treated reasonably by his jailors.[5]

In 2008 while on another of my annual visits to libraries and archives I continued my search for material on the last days of the Tirot Sing. The National Archives of India located in Janpath, New Delhi, is the repository of official manuscripts and publications of the Government of India. It traces its records to early British rule. I had read in other repositories and had just two days of research in this archive. Fortunately a friend and archivist, Mr. Mehra, helped me quickly locate the index of the Foreign Department Proceedings from 1833 through to 1835. From the index I was able to call for documents.

A letter dated 12 February 1833 from T.C. Robertson, the Agent to the Governor General North Eastern Frontier to the Chief Secretary, Government of Bengal informs of the surrender of the Khasi *Syiem* and that he was being removed from the vicinity of the hills. Robertson informs that he had requested the Magistrate of Dacca to detain the *Syiem* in his jail as a state prisoner. The same letter enclosed a copy of a letter from Lt. Henry Inglis of the Sylhet Light Infantry to Captain Lister; Commandant of the Sylhet Light Infantry dated 14 January 1833. This is an important letter for the information which it highlighted. It informs that Tirot Sing come on the date of his surrender with a party of thirty swordsmen and eleven musketmen ..."I remarked to Jeet Roy that this was a breach of the promise he had made, when he replied that it would not have been respectful for his master to come without a small retinue and that this was to show that he was not made captive but surrendered."

A British officer notes that Tirot Sing arrived Dacca with nothing but a blanket and a plate or two. He was visited by Middleton, the Commissioner of Circuit for the District of Dacca soon after his arrival. Not long after a letter was addressed to Middleton from the Secretary of the Bengal Government that the Governor- General agreed with the opinion "that it will be objectionable to confine Tirot Sing to the common jail if a more appropriate place of safe custody can conveniently be procured for him." On 16 April 1833 Middleton informed Fort William, Calcutta that he had personally inspected two other houses before recommending a third house which was proposed to be rented for Tirot Sing. "It is a lower one with as many apartments as can be wanted and within the compound, which is enclosed by a wall having only one gateway, secured by a massive door."

The house was guarded by a *Burkandaz* of the Dacca *Kotwally* who was regularly relieved. The *Darogah* of the *Thana* was instructed to exercise a general superintendence. Mention is made that the house was located at Girdkillah of old Dacca city. Official records are silent about the life of Tirot Sing from the summer of 1833 to the record of his death. He was first imprisoned in the common jail, later he was placed under house arrest and by the year of his death was given the liberty to move around Dacca. What eventually caused the death of the man is not certain though tradition says it was the outcome of a stomach ailment.

A note in the index of the Foreign Political Proceedings at the National Archives informs researchers that records after 21 November 1834 are preserved in the West Bengal Archives, Kolkata, which is where I next went. Here too I received assistance from the Director of Archives and his Deputy. It was not long before a voluminous index of the Bengal Political Consultations 1835, volume 2, was placed on my table. It was only a matter of minutes before I located the relevant consultation. The volume was requisitioned and placed on the table. Sifting through the pages I read the same letters I had many years earlier uncovered in the India Office Library and Records, London, of Scott the Officiating Magistrate of the Foujdary Adalat, Zillah of Dacca writing to J. Lewis the Officiating Commissioner of Circuit, Dacca, informing the death of Tirot Sing.

Many winters ago, I was in Dacca to research on a book. The house I resided in was in old Dacca. Close by was the palatial structure of the former *Nawab* of Dacca and an old graveyard. Not far away was the library of the Asiatic Society of Bangladesh. Much further was the Bangladesh State Archives. No clues came from my queries of Tirot Sing at these places other than some information on the prison records of that time, the names of prisoners and the daily allowance they received. Tirot Sing was not among the names of prisoners.

It was in Kolkata in February 2008, I learned when closing the record of the death of the Khasi *Syiem*, why there was nothing more on Tirot Sing in official files. A note below the letters mentioned above was recorded "No order". Nonetheless the search goes on for more on Tirot Sing.

Conclusion

For too long have the Khasis believed and made others believe, that Tirot Sing died in jail. Though it is historically correct to say that he remained a prisoner from the time of his surrender in January 1833 to his death in July 1835, the truth of his death is that Tirot Sing did not die in prison, at least not in a prison cell. He must have longed to return to the hills as a drama on the man depicts; he could not, due to the nature of his confinement. He certainly had a sad life, away from his people and family, and hastened by sickness that caused his untimely death.

The nation observed the 150th anniversary of the 1857 uprising some years ago. Those who challenged British imperialism were remembered. The nation by commemorating the 125th anniversary of the establishment of the Indian National Congress, has given society another opportunity to recollect the effort of those such as Tirot Sing who fought for their rights and land. Some interest is understood to be growing in Dhaka on the number of Indian rulers including Tirot Sing who were placed under house arrest in the old city.

This research on the date of death of Tirot Sing lay to rest the conflicting dates on the date of death of the *Syiem*. The Khasis of our times however, have not accepted the truth of the status of the man as a state prisoner and the rights he enjoyed as an imprisoned India ruler. They continue to harp on the wretched end of the *Syiem* even though the British had recognized him as a former ruler and treaded him kindly. Most myths become popular when they appeal to sentiments. They need not necessarily be based on historical truths. With the information we now have on the last days of the Khasi chief, the monuments with inscription such as those at Mairang and Shillong should be reworded and histories should be rewritten. Efforts should be taken to give Tirot Sing the right and more respectable account of his last days and demise.

Notes and References

*David R. Syiemlieh, 'Call of Freedom from the Hills: Tirot Sing and his significance in the Freedom Struggle', Sanjoy Hazarika (ed.), *Little Known Fighters against the Raj: Figures from Meghalaya*, Centre For North East Studies And Policy Research, Jamia Millia Islamia, New Delhi, 2014, pp.31-42.

1. H. K. Barpujari, Presidential Address, *Proceedings of the NEIHA*, First session, Shillong., 1980, p. 12.

2. For a general but comprehensive account of British expansion into North East India expansion read H. K. Barpujari, *Problem of the Hill tribes North East Frontier*, volumes 1-3, reprinted NEHU, 1998; H. K. Barpujari, *A Comprehensive History of Assam*, volumes 1-V, Guwahati, 1993-1994. For the colonial connection with the Khasis read Hamlet Bareh, *The History and Culture of the Khasi People*, Calcutta, 1967; David R. Syiemlieh, *British administration in Meghalaya: Policy and Pattern*, Delhi, 1989 and Helen Giri, *Khasis under British Rule (1824-1947)*, Regency Publications, New Delhi 1991.

3. Alexander Mackenzie, *History of the Relations of the Government with the Hill Tribes of North East-East Frontier of Bengal*, Calcutta, 1884, reprinted *The North –East Frontier of India*, Mittal Publications, Delhi, 1979, p. 222; R. B. Pemberton, *Report on the Eastern Frontier of British India*, DHAS, Guwahati, 1966, p. 232.

4.   "New Light on Tirot Singh: His Last Days and Demise", *The NEHU Journal of Social Sciences and Humanities*, Vol. V, No. 4, October-December 1987, pp. 27-30.

5.   "Two Perceptions of Tirot Sing", *Proceedings of the NEIHA*, Ninth Session, Gauhati, 1988, Shillong, 1989,pp. 262-273.

11

The Last Days of David Scott

Introduction

David Scott is too well known to those interested in the history of North East India to require any account of his work in the region. What is not quite known is the man's last days. There are two biographies on Scott. Adam White's, *A Memoir of the Late David Scott* was written and published soon after Scott's death. Nirode K. Barooah's, *David Scott in North-East India: A Study of British Paternalism* is a much more detailed study having originally been a doctoral thesis. White's book provides much information on Scott and his health conditions. These come in the form of correspondence of Scott and other British officials. Barooah had access to the manuscripts on Scott in Nottingham University but was not inclined to include this in the biography. This essay will recall Scott's dedication to work, as a result of which his health failed; his attempts at self medication; the concern at Calcutta of losing Scott; his last words; his death and arriving at the cause of the early demise. The essay will conclude with a note on the Scott monument.

Khasi-Jaintia Connection

The last years of Scott's life were largely spent in the Khasi hills. He first had an opportunity to reside in these hills when he first traversed through them in 1824 to open negotiations with Ram Singh, the Jaintia *Raja* and the Khasi *Syiems* of Nongkhlaw and Sohra, for the construction of a road through the hills to connect Sylhet and Assam. Delighted with its salubrious climate he brought it to the notice of Government as offering very desirable situations for sanatoria stations for Europeans. He was able to secure permission from Tirot Sing for the construction of a house at Nongkhlaw "to eat the Europe air."[1] After a year's residence there he was affected by a severe sickness that affected other Europeans and Khasis.[2] Nongkhlaw was therefore ruled out for further development as a health resort for Europeans though Scott continued to be based there. Attention was then directed to start a sanatoria station and cantonment for European troops at Cherrapunji.

Scott was convinced "of the immense advantage that the European forces would derive from being cantoned in this quarter. If the recruits were at once brought up here they would have no opportunity of acquiring idle dissolute habits" he wrote.[3] He looked forward to the development of the Khasi, Nilgiri and Simla hills as sites for establishing colonies, in which a race of hardy European soldiers might be reared, capable of defending British interests and territories in India. This view was speculated, wrote his first biographer "in view of the possibility of our maritime supremacy being endangered, thereby cutting off supply of recruits; although doubtless other contingencies entered into his calculation.[4] George Swinton, Chief Secretary of the Bengal Government was as enthusiastic as Scott to establish European cantonments in the Khasi Hills. The Governor-General William Bentinck's own views on Scott's early plans for the development of a military colony in Cherrapunji seemed to him "wild and impracticable, but I have no doubt the Europeans will be induced to settle in those hills, if further trial confirms the belief entertained of their healthiness."[5]

Scott escaped death many a time. He survived the killing of Nongkhlaw of 4-5 April 1829 as he had left for Cherrapunji a few days earlier in connection with the construction of the road. Many other attempts were made to take his life. What saved Scott from a more untimely death were the pre-monitions of danger that made him take precaution in never staying in a place where he would suffer the fate of Bedingfield and Burlton. Scott was always to be seen with his double barreled guns and swords.[6] The suppression of the Khasi uprising, the problems of administering newly acquired Assam, the Garo Hills with bordering Goalpara and Mymensingh and Sylhet took much of Scott's time. As if this was not enough work he busied himself in encouraging missionaries to set up schools and missions while he himself planned the economic development of the hills and the expansion of the British base at Cherrapunji. Fruit trees and vegetables of various description were written for from Calcutta; cattle to improve the local breed, sheep to be reared at Barduar, roads were planned and bridges were to be constructed. These experiments and plans were made not only to improve the lot of the Khasis but more important from a political point of view was his intention that a large European colony of around ten thousand persons could live in these hills permanently not very different from what they could be used to in England.[7]

Decline in Health

Such a, wide variety of responsibility and interests involved much travel. Rough horse rides up and down the hills into Assam and Sylhet to attend to his official duties as Agent to the Governor-General on the North East frontier of Bengal and as Commissioner of Revenue and Circuit of Assam, North-East Rangpur, Sherpur and Sylhet were a tiring affair, especially so as Scott was tall and inclined to corpulence.[8] His bulky figure however did not deter him from his duties nor did the sultry climate of the plains. Being a bachelor he had no attachment for home, which for him was where he attended to duty. Consequently he was over working himself.

Scott had a heart disease which required him to sleep in a sitting position and frequently prevented him from sleeping at all. Although he could scarcely walk from the palpitations of his heart, which he seldom attempted, he continued to keep a busy official schedule with long and difficult journeys. He would often begin work at sunrise and would remain in the *kutcherry* until sunset, under the warmth of his blanket for he never used fires.[9] Sometime in 1830 he wrote from Nongkhlaw to Dr. Lamb in Cherrapunji seeking his advice whether he should proceed to Calcutta for a sea journey or to go to a place of colder climate. But, this incredibly conscientious officer answered his own question "that the urgency of my business there is not however, so great as to induce me to go."[10]

By June 1831, Scott's conditions worsened. In a letter to his friend George Swinton, Chief Secretary of Government, Scott wrote:[11]

> I arrived here (Nongkhlaw) the day before yesterday in hope of obtaining some relief from the distressing symptoms I have been laboring under for the last four or five weeks. During this time a great change has taken place, I fear for the worst and something of consequence, must, I apprehend have happened to the structure of the heart itself, the motion being now very different from what it was, and the throbbing consisting rather in a general movement of the whole body than in the direct beating of the heart. I cannot sleep, I am troubled with frequent sickness at stomach, and am exhausted with the least exertion.

All these symptoms appeared about 10 May after he had taken too large a dose of hydrocyanic acid which he had hoped would cure the stomach disorder.[12] Two days later he wrote a post- script that no change for the better had taken place and that he thought of going over to Cherrapunji although he feared that the doctors there could do little.[13] Some weeks later, still at Nongkhaw, he wrote to Swinton again that he had a good appetite and was able to sleep but complained of a terrible swelling in his legs.[14] His friends at Calcutta advised him to leave the hills and take a sea voyage but Scott intended to remain at Nongkhlaw and consider the sea voyage until matters were settled in Assam. By then he was unable to hold sessions at Sylhet and Goalpara for he was bedridden.[15]

An extract of a letter from Scott to Lt. Hamilton Vetch dated 2 June 1831 informs the young officer that he was going to the hills 'once more'. He was finding it difficult breathing, sleeping and eating, and that Vetch should not be surprised "if you hear that I have taken a place beside our poor friends on Ostrich Hill."[16] Swinton was sad that someday Scott, his friend and classmate at Fort William College, would be no more. He and Dr. Nicholson at Calcutta were expecting Scott's arrival in the Presidency, but so too was Dr. Rhodes at Cherrapunji. The torrential rains that the hills were then experiencing, it was believed, must have prevented Scott from moving from Nongkhlaw.[17] Dr. Nicholson, therefore, proposed that Scott should not leave the hills. Swinton who must have wished to see his friend hoped "that his valuable life and services will not be lost to us."[18] We do not know when Scott moved from Nongkhlaw to Cherrapunji but that he did do so is no doubt for Swinton had heard from Cherraunji that Scott had left the hills very ill.[19] The journey from his cottage at Nongkhlaw to Cherrapunji must have weakened the man as a result of which he could not descend the hills towards Syhet and had to spend his last days in Cherrapunji.

Two of Scott's last letters were written from Cherrapunji. In one of these last lines he informed Government that he was unable to perform his official duties.[20] The other to a near relation residing in Calcutta and dated 14 August 1831 was written in a tone of cheerfulness in every way calculated to allay the apprehensions that his friends feared.[21] His last words so typical of the man were, "I wish you gentleman," he told Colonel Watson, Dr. Rhodes and Lt. Day, "to bear witness to Government that I am no longer able to conduct that affairs of the country."[22]

Death

David Scott passed away a few minutes after 6 o'clock in the morning of 20 August 1831 at the age of 45 years. He was buried on the evening of his death on a hillock in the British settlement in Cherrapunji by the side of Ensign Brodie. The funeral was attended by all at the station and with full military honours. Dr. Rhodes conducted a post mortem before the burial to find out the exact cause of the death as the case of Scott was a peculiar one that had excited considerable interest. Rhodes recollected that Scott had himself wished such an examination be made. The body was excessively yellow, the legs and scrotum much distended the cellular membranes of the chest and neck very much discoloured with extravagated blood especially about the neck and the hands particularly the fingers were nearly black. Upon opening the body Rhodes found the heart larger than usual and very pale, weighing 1 pound 7 ounces when freed from all blood. He examined the aorta down to its bifurcation at the loins, but found no enlargement. On removing the heart, he found on dissection, the right auricle of the usual size and the tricuspid. valve healthy, as were also the right ventricle and semilunar values of the pulmonary artery. The left auricle was enlarged, as was the ventricle, and the walls of the latter much thickened.[23]

Rhodes' post-mortem report continues:[24]

I now come to the seat of the disease, and cause of the patient's death. On attempting to pass my fingers along the aorta, I found it obstructed at the semilunar by a bony substance and on examination found all the values of the aorta assified, and the vessel itself almost totally blocked up by a honey-comb like bony substance, leaving a space of about 1/4 of an inch in length, and the width of a probe for the passage of blood.

The diagram to the above account further mentioned that all traces of the semilunar valves were lost. In cutting into the abdomen about two pints of bloody serum was discharged, part of which had escaped from the chest through an opening made in the diaphragm. The liver was enormously enlarged, of a very black colour and when cut into showed itself loaded with dark blood and bile. The head was not examined. These conditions therefore accounted for Scott's death which had been so perceptible for years. Rhodes who had treated Scott for many months was surprising how he could have lived so long.[25]

Conclusion

Few early British administrators have received as much attention as Scott's death did for him. *The India Gazette* in an official notice of his death said that he was "a most zealous, faithful and intelligent public servant.[26] One of his close associates from

Rangpur where he has earlier been posted, wrote in the *Bengal Hurkara* and *Bengal Chronicle* that "to the sharpest and brightest intellect added a winning simplicity of character, combined with a benignity so deep and searching,. to the great extent of his charity and the singleness of his purpose in everything he undertook.[27] His successor as Agent to the Governor-General, William Cracroft, who had known Scott since their student days at Fort William College, Calcutta, hoped he would be no unworthy successor of "so bright a character and one possessed of talents and judgment."[28] Government authorized the construction of a simple monument to be erected over his grave in token of its regret of the loss of David Scott.[29] Close to the Circuit House at Cherrapunji still stands this monument, one of the very few memorials built for Company servants on which is inscribed:[30]

> This monument is erected by order of the Supreme Government as a public and lasting record of its consideration of the personal character of the deceased and of its estimation of the eminent services rendered by him in the administration of the extensive territory committed to his charge. By his demise Government has been. deprived of a most zealous, able, and intelligent servant, whose loss it deeply laments, while his name will long be held in grateful remembrance and veneration by the native population, to whom he was justly endeared, by his impartial dispensation of justice, his kind and conciliatory manners, and his constant endeavours to promote their happiness and welfare.

An official, appreciative of Scott's service wrote he was "indeed a second Cleveland." By far the greatest tribute to Scott came from Alexander Mackenzie who wrote of him in his official history of British imperialism:[31]

> Had the scene of his life's labours been in North-West or Central India, where the great problem of empire was being worked out, instead of amid the obscure jungles of Assam, he would occupy a place in history by the side of Malcolm, Elphinstone and Metcalfe.

Notes and References

*David R. Syiemlieh, 'The Last days of David Scott,' *Proceedings of the North East India History Association*, Fifth Session, Aizawl, 1984, Shillong, 1985, pp.107-113.

1.   'Chirra Punji,' *Asiatic Journal*, Vol.XXI, 1836, p.17; A White, *A Memoir of the Late David Scott*, Calcutta, 1832, p. 37.

2.   R. B. Pemberton, *Report on the Eastern Frontier of British India*, DHAS, Gauhati, 1966, p. 255.

3.   National Library of Scotland, Dalhousie Muniments, GD45/5/48, David Scott to George Swinton, 27 October 1830.

4.   A. White, op. cit., p. 51.

5.   C. H. Philips (ed.), *The Correspondence of Lord William Cavendish Bentinck*, Vol. I, Oxford University Press, Oxford, 1977, No. 237, p.500. Scott had submitted a detailed suggestion for the establishment of a cantonment of 300 to 400 European soldiers in the Khasi hills. He wrote to Government for a money advance to establish the colony which could be paid back in four years from the profit of dairy and agricultural business in the hills. Nottingham University Library Archives, Bentinck Papers in the Portland Collection, hereafter Bentinck Papers, PWJF 2791/1, David Scott to George Swinton, 21 July 1830.For details of this plan read Nirode K. Barooah, *David Scott in North-East India 1802-1831 A Study in British Paternalism*, Munshiram Manorahlal, New Delhi, 1970, 211-229. The 'Cherra Punji experiment' for a sanatorium and cantonment did not take shape. By the mid 1830s attention was drawn towards Darjeeling, recently acquired by the British. For this shift in policy read David R. Syiemlieh, "The Cherrapunji Experiment (1829-1834)", *Proceedings of the NEIHA*, Fourth Sesssion, Barapani, 1983, Shillong, 1984,pp.116-123; and "Cherrapunji Versus Darjeeling: The search for a Sanatorium for the Lower Provinces," *Proceedings of the NEIHA*, Sixth Session, Agartala, 1985, Shillong, 1986, pp.219-224.

6.   A. White, op. cit., pp. 40-49.

7.   *Ibid.*; PWJF 2791/1, David Scott to George Swinton,21 July 1830;PWJF 2820/IV, David Scott to George Swinton, 2 June 1831; National Archives of India, Foreign Political Consultations, 10 February 1826 David Scott to George Swinton,3 January 1826.

8.   A. White, op. cit., Appendix 40,p. 133.

9.   *Ibid.*, Appendix 5, p. 78-79.

10. A. White, op.cit., Appendix 17, pp.104-105.

11. Bentinck Papers, PWJF 2811/XXV, David Scott to George Swinton,12 June 1831.

12. *Ibid.*

13. *Ibid.*

14. Bentinck Papers, PWJF 2811/XIX, David Scott to George Swinton,24 July 1831.

15. *Ibid.*

16. A. White, op. cit., Appendix 30, p.118. 'Ostrich Hill', a location in Cherrapunji where Lt. Beddingfield was buried.

17. Bentinck Papers, PWJF 2811/XXII, George Swinton to R. Benson, Military Secretary to William Bentinck,1 June 1831; PWJF/2811,George Swinton to R. Benson, 14 July 1831; PWJF/ XXVI, George Swinton to R. Benson, 9 July 1831.

18. Bentinck Papers, PWJF 2781/XXXV, George Swinton to R. Benson, 26 July 1831.

19. Bentinck Papers, PWJF 2811/XXXVI, George Swinton to R. Benson, 28 July 1831.

20. West Bengal State Archives, Judicial and Criminal Proceedings, 9 August 1831, David Scott to James Thomson, 25 July 1831.

21. A. White, op. cit., Appendix 37, pp.126-128.

22. *Ibid.*, Appendix 43,pp. 136-137.

23. Bentinck Papers, PWJF 2811/XIII, Dr. Rhodes to George Swinton, 21 August 1831.

24. *Ibid.*

25. *Ibid.* An official report on Scott's death mentions that it was due to "ossification of the heart." A. White, Appendix 43, pp.136. H.T. Princep, Secretary to the Governor-General to George Swinton (then in Simla), 20 September 1831.

26. A. White, op. cit., Appendix 38, pp. 126-127.

27. *Ibid.*, Appendix 39, pp. 127-128.

28. Bentinck Papers, PWJF 2811/IX, William Cracroft to George Swinton, 3 September 1831.

29. A. White, op. cit., Appendix 43, p.136.

30. The words on the plaque on the Scott Monument in Cherrapunji was taken from a letter from H.T.Princep to George Swinson, Bentinck Papers, PWJF 2811/XVI, George Swinson to R. Benson, 30 August 1831; A. White. op. cit., Appendix 43, pp.137-138.

31. Alexander Mackenzie, *History of the Relations of the Government with the Hill Tribes of the North-East Frontier of Bengal*, Calcutta, 1884, reprinted *The North-East Frontier of India*, Mittal Publications, Delhi, 1979, p.5 footnote.

12

Remembering Thomas Jones

*" If I should go I could be instrumental to
Make known the way to salvation to
thousands that should never hear of it
should I not go."*

Thomas Jones: 29 August 1839

We are so often caught up with the present and what the future has in store that we forget the past and little realize what the past makes of the present. In jotting this remembrance of Thomas Jones I pay tribute to this first Welsh Missionary to the Khasi-Jaintias and two other Thomas Jones enthusiasts; Nigel Jenkins whose *Gwalia in Khasia*[1] first gave us an insight into the life of the man, and Jones' great grandson Dr. Andrew May, who came to Shillong some years ago to trace his own roots and researched on Thomas Jones. Dr. Andrew May is professor of history at Melbourne University, Australia. He had intended to come to Shillong to take part in the Thomas Jones, birth bi-centenary celebrations. He informed that he could not make it to our hills but that he is near completing the writing of a book on his great-grandfather, Reverend Thomas Jones, founder of the Presbyterian Church in the Khasi-Jaintia Hills.[2]

How did I become a Thomas Jones enthusiast? It began many years ago, when after completing my PhD at NEHU, I spent a month in Calcutta browsing through archives and libraries. It was January 1986 I recall when I located the grave of Thomas Jones in the Scottish Cemetery on Karaya Road. Over the years and with more research on the man in archives in London and Calcutta, I was able to piece together part of the life of this fascinating personality.[3] Several other scholar published references to Jones' efforts in journals and newspapers recollecting the story of Thomas Jones and keeping alive in our times the saga of the man.[4]

Whenever I go to Calcutta, the "city of palaces", I find my way to the Scottish Cemetery on Karaya Road and pay my respects to a man whose work was so little appreciated. On 24th of January 2010 on the bi-centenary of his birth, I, like the owners of the mill in Berriew, Wales, where Thomas Jones grew, placed a bunch of daffodils in front of a picture of the man. He will be remembered here in the Khasi-Jaintia Hills, in Wales, in Scotland, in Australia where his descendants are settled and perhaps in many other part of this region. It was fortunate that the bi-centenary of his birth happened to be a Sunday. He was thankfully mentioned in prayers and sermons. And often at these functions the song *Ri Khasi Ri Khasi* is sung, which many Khasis may not know was put to verse and song by a missionary and which follows the tune of the Welsh national anthem *Hen Wlad Fy Nhadau* (*Land of My Fathers*) and the Breton national anthem *Bro Gozh ma Zadoù* (*Old Land of My Fathers*).

The Khasi Hills Presbyterian Church has done well to place a marble cover to the grave and a plaque remembering the "Founding Father of the Khasi Alphabet and Literature and the Pioneer of the Welsh Presbyterian Mission in the Khasi Hills." The members of the Khasi congregation of a church in Calcutta annually lay wreaths on the grave on 16 September, the death anniversary of Thomas Jones. A book was published last year on Jones, a college is named after him, the Thomas Jones School of Mission bears his name and several schools are named after the man who gave the Khasis the A,B,K,D and more. Primary School teachers of the Khasi Hills honour the missionary on his death anniversary. The Khasi Jaintia Presbyterian Assembly has drawn up a programme to commemorate the man who started the mission in these hills. As the celebrations unfold in the months ahead we would have done much to remember this Welshman, missionary and fine human being.

Born on 24 January 1810, in Tanyffrid, Montgomeryshire, Wales, Jones was one among ten children of Edward and Mary Jones. After living in Liverpool for six years the family returned to Berriew and settled in the Llifior mill where the young Jones learned a trade as miller and joiner. Jones later studied theology at the Bala Calvinistic Methodist College after which he volunteered to be sent out as a missionary of the London Mission Society.[5] He was destined not to go to Africa where the LMS wanted to send him. In January 1840 the Welsh Presbyterians broke off from the LMS and set up their own mission society. Thomas Jones was their first missionary.

Jones married shortly before sailing for India. On 4 November 1840 some 600 supporters of the new foreign mission assembled at the Rose Place Church, Liverpool to bid the young missionary and his wife *bon voyage*. On that day Jones signed a document of affirmation in Welsh to hold fast to his church's doctrine, the church structure and disciplinary rules.[6]

Setting out for Cherrapunji as it was then called, with his wife Anne, soon after that affirmation, the Jones arrived Sohra in the monsoon of 1841. Before Jones' arrival, Sohra and the region around had heard the teachings of the English Baptist missionary Alexander Lish. After him came the LMS missionary Jacob Tomlin who spent some months in Sohra. The early decades of the eighteenth century witnessed a flourish of mission societies in the West. Almost all were to have some connection with the spread of Christianity in Africa, Asia and further east.

Despite the difficulties of climate, language and funds, Thomas Jones set up the mission at Nongsawlia, Sohra. By early the following year he opened the first of three schools in Sohra, Mawmluh and Mawsmai. That same year he had worked on and returned from Calcutta with the first Khasi primer in the Roman script. Despite criticism for the decision this was the start of the medium for instruction and the very first beginnings of Khasi literature. The official policy then was to encourage the use of Bengali for the Khasi schools and nascent literature. The success of this decision to use the Roman script was in time to become the vehicle by which other mission societies reduced to writing the languages of the numerous other tribes in the North East. In this sense therefore Jones' impact goes far beyond the Khasi Hills. From Sohra the Presbyterian church spread its influence into Sylhet, Cachar, Mizoram, the hills of Manipur and the North Cachar Hills. Jones translated the Lord's Prayer into Khasi; he translated the Welsh catechism, he published *Ka Kitab Nongialam* and worked on the translation of the *Gospel of Matthew*, which was published in 1846.[7] A visit to Jowai in 1845 was followed by the start of the mission among the Pnars.

Thomas Jones worked in Sohra for eighteen months before other missionaries arrived. Jones had the distinction of never having converted anyone but the church was set on foundations he laid. He was more interested in improving the economic lot of the people. Differences with Reverend William Lewis on the purpose of mission and the Mission Board at home which was not particularly pleased with his approach resulted in Jones leaving his missionary calling. If this was a concern for the young man, his wife Anne died in August 1845, shortly after the birth of their third child. The child too passed away a few days later. A year later Thomas Jones married Emma Jane Cattell on 16 September 1846. The "injudicious marriage"[8] as the official history of the mission reports and the business endeavours of the man further distanced Thomas Jones from the Mission Directors. The connection was severed in 1847.[9] Sadly though Jones' business brought him the ire of Henry Inglis, former Assistant Political Agent, Cherra Political Agency. Inglis had a commanding control over the business activities in the Khasi Hills. Harried by his countryman, (would it be correct to say this because Henry Inglis was born in Sylhet, lived his entire life in and around these parts and died in Sohra - he had never set foot on the British Isles!) and weak after an illness, Jones died of what appeared to have been malaria, in Calcutta on 16 September 1849. He was buried the following day.

His son Thomas Cattell Jones was born posthumously. Raised in Sylhet, Thomas Cattell Jones' son grew up in Weybridge, England, was trained in Edinburgh in the medical profession and returned to serve in Sylhet.

It has taken the church a long time to honour the man so long forgotten and little spoken of. Did Thomas Jones deserve this ignominy? The church in our hills only awoke to honour the man but recently. This has come timely and not too late for Thomas Jones now finds a respectable place in Khasi history. The church in Wales too has changed its views on their first missionary to the Khasis, for what Jones started was to become the largest overseas venture of the Welsh church. Time heals; Thomas Jones is fondly remembered today for what he had set out "to make known the way to salvation to thousands that should never hear of it should I not go" and more.[10]

Notes and References

This essay was first published in *The Shillong Times*, 24 January 2010 on the 200th anniversary of Thomas Jones' birth.

1. Nigel Jenkins, *Gwalia in Khasis: The biggest overseas venture ever sustained by the Welsh*, Gomer, Dyfed, 1995.

2. Andrew May, *Welsh missionaries and British imperialism: The empire of clouds in north-east India,* Manchester University Press, 2012.

3. David R. Syiemlieh, 'Thomas Jones' Injudicious Marriage*?*' *Proceedings of the North East India History Association*, Fifteenth Session, Doimukh, 1994; pp.240-244; David R. Syiemlieh, 'More on the Thomas Jones Saga,' *Proceedings of the North East India History Association,* 1999, pp.200-203; David R. Syiemlieh, 'Thomas Jones' Application and Affirmation to become a Missionary to the Khasis,' *Proceedings of the North East India History Association*, Twenty- eighth Session, Goalpara, 2007, pp. 301-306.

4. Andrew May, 'The promise of a book; Missionaries and native evangelists in north-east India', *Missionaries, indigenous peoples and cultural exchange*, Sussex Academic Press, 2010,pp. 81-95; 'Homo in nubibus: Altitude, Colonisation and Political Order in the Khasi Hills of Northeast India,' *Journal of Imperial and Commonwealth History*, vol. 42, 2014, pp.41-60; S. S. Majaw, *U Thomas Jones bad Ka Pyrthei Saitpohpen*, Shillong, 2011.

5. D. Ben Rees (ed.), 'Thomas Jones (1810-1849),' *Vehicles of Grace and Hope: Welsh Missionaries in India 1800-1970,* William Carey Library, 2002, p.101.

6. David R. Syiemlieh, 'Thomas Jones' Application and Affirmation to Become a Missionary to the Khasis,' *Proceedings of the North East India History Association*, Twenty-eighth Session, Goalpara, 2007, pp. 301-306

7. Two translations by Thomas Jones of the Gospel of St. Matthew (1846) and the four Gospels and the Acts of the Apostles (1856) published by the Calcutta Auxiliary Bible Society are in the British Library, London.

8. David R. Syiemlieh, "Thomas Jones' Injudicious Marriage*?*" *Proceedings of the North East India History Association,* Fifteenth Session, Doimukh, 1994; pp. 240-244. I recall reading an essay by R. T. Rymbai which came to him by oral tradition and mentioned Thomas Jones' marriage to a 'Kong Kuttel.' This came before the more recent research on the life of the man. It is striking that there is a resemblance and closeness of the name of Emma Cattell and that mentioned by R. T. Rymbai.

9. Andrew May provides detail of the closure of the connection. Read *Welsh missionaries and British imperialism: The empire of clouds in north-east India,* Manchester University Press, 2012, pp.190-193.

10. This last line and the quotation at the start of this essay come from the application dated 29 August 1839, made by Thomas Jones to the London Missionary Society. London Missionary Society Papers, School of Oriental and African Studies, University of London.

13

Among Many Writers: Contribution of Homiwell Lyngdoh to Khasi History

Biographical Sketch

Homiwell Lyngdoh came from the Nonglait clan of Mawklot village close to upper Shillong. His mother was Thangdamon Nonglait. His father Bir Singh Nongkhlaw was a *Myntri* of Mylliem *hima*. Born on 26 February 1877[1] he was among the first generation of Khasis to have received an education. After his schooling in the mission school in Shillong he proceeded to Calcutta where after a year pursuing the first arts class, he joined the Calcutta Medical College. He passed the Licentiate in Medicine and Surgery in 1903.Homiwell Lyngdoh had an illustrious career that took him to Gauhati, Dhubri, Chittagong, Imphal, Nowgong and Silchar. He was posted as Civil Surgeon in Shillong for five years during which time it is said he got involved in activities which were to make his name remembered as a writer. He also served for several years on a temporary commission in the Indian Medical Service. Government awarded Dr. Lyngdoh the medal of the Order of the British Empire. In his retirement from government service, Homiwell Lyngdoh involved himself in the political activities in the Khasi hills with the revival of the Federation of Khasi states, the formation of the Khasi Jaintia Political Association and the concerns in these hills with the transfer of power from Britain to independent India. Homiwell Lyngdoh passed away on 22 April 1958.[2]

In the select bibliography in English and Khasi of references used in Hamlet Bareh's *The History of Culture of the Khasi People*, the author notes that until 1967-68, "nothing on the history and culture of the Khasis came out. Thereupon my effort to produce it"[3]. It may be true that nothing substantial and comprehensive had been written in English before Bareh's *History*, though his interpretation of events had to depend on existing published histories many of which were in the Khasi language including the publications of Homiwell Lyngdoh. A close look at his Khasi sources and a search for what had preceded and came after this first comprehensive history of the Khasis reveals a fair interest among the Khasis society in their history, the nation and even world events. That the Khasis had a sense of history is reflected in their oral tradition and folklore. That today these traditions are being lost out with the many influences on the society and the lack of appreciation of how the Khasis came to be till the present, is sadly reflected in their lack of interest in the past.

Early Literature

Thomas Jones, the first Welsh missionary to the Khasis started the use in 1842 of Roman Script for the Khasi language. In the first flush of literary works a combination of primers, Christian literature and translations were published. Beginning with a translation of Rhodd Mamm (1842), a popular Welsh story, Jones worked on *Ca Citap Nyngkong Ban Hicai Poole Ci Citin Cassi* (*First Khasi Reader*, (1842), *Oo Nonghicai* (*The Instructor*, 1843), and *Ca Gospel Jong Oo Mathiaos,* 1847). By the time Rev. William Lewis who came after Thomas Jones, tried his hand in the new expression some standardization had taken place. He added a small catechism, a hymn book (1850) and a translation of the Four Gospels and the Acts of the Apostles (1856).[4] In 1855, Rev William Pryse's *Khasi Grammar* was published. In the Prefatory Note to the vocabulary, Pryse takes a stand that the Khasis, "a small and uninfluential tribe will not be able to retain characters different from those of the larger natives of the plains which surround their hills." He continued:[5]

> Should the Khasia tribe be ever brought under the influence of education, civilization, and commercial intercourse, the Bengali character must supplant the Roman at a not very distant day.

Pryse may have been influenced in this opinion as he had started work among the Bengalis of Sylhet before going up into the hills. Perhaps a not too encouraging start was made with the setting up of three schools at Mawmluh, Mawsmai and Cherrapunji where Khasi was taught with the British official policy of encouraging the use of the Bengali script.[6] The debate of which script to use was not a long one. By the 1860's-1870's the decision of the Welsh Mission to use the Roman Script prevailed.[7] It was followed by an outpouring of literature that stimulated literary creativity in prose, poetry and drama; the establishment of Ri Khasi Press (1896) and monthly journals *U Khasi Mynta* (1896), *U Nongphira* (1903), *Nongkit Khubor* 1889), replaced in 1902 by *U Nongialam Kristan.* It was not long before Calcutta University, where many Khasis first went for higher education, gave recognition of the development of the language in its many expressions when it decided in 1902 that Khasi could be included in the use of subjects for the Entrance Examination.[8]

Before and by the turn of the century fragments of Khasi history were appearing in Alexander Mackenzie, *History of the Relations of the Government with the Tribes of the North-East Frontier of Bengal* (1884); W.W. Hunter, *A Statistical*

Account of Assam, (1879); C.U. Aitchison, *A Collection of Treatise, Engagements and Sanads* (1892); (1909); J.H. Morris, *The History of the Welsh Calvinistic Methodists' Foreign Mission* (1910); and D. Herbert, *Report on Succession to Siemships in the Khasi Hills* (1902). These and other material in the *Journal of the Asiatic Society of Bengal*, reprints of A.J.M. Mills, *Report on the Khasi and Jaintia Hills 1853* (1901), and W.J. Allen, *Report on the Administration on Cossyah and Jynteah Hill Territory, 1858* (reprinted 1903), would have enabled B.K. Sarma Roy to write *Ka Histori Jong Ka Ri Khasi,* Shillong, 1914. The book of 104 pages has chapters on the Khasi states, a chapter on neighbouring Manipur and one on the Syiemlieh clan.[9] From the title of this first history in Khasi, its subject was the Khasi people. As an observer of the society B.K. Sarma-Roy's work becomes important as a start and reference point.

First Detailed History

Those who wrote on the early history of the Khasis were not trained historians. They were largely administrators and missionaries and the first of the educated Khasis, many of whom had attended school and college in Calcutta. Homiwell Lyngdoh's first effort in writing history was *Ka Pom Blang bad Thang Syiem Sohra* (1928). Early in 1925 he had closely observed the elaborate ceremonies related with the cremation of *Syiem* Roba Singh[10] of Sohra and his succession by Join Manik Syiem. In May that same year he recorded the religious ritual of the Khyrim state. The death rituals and process of the embalming of the Sohra *Syiems* is described in detail. Similarly his note on the ceremonies at Smit in Khyrim *hima* were the most detailed description of this significant tradition in the Khasi order of state and religion. This very significant study was followed by *A Short Account of Cherrapunji* (1933) which was published on the occasion of the visit of the Governor-General Lord Willington to Sohra in October 1933.

Homiwell Lyngdoh's connection with the Presbyterian church of which he was an elder did not deter him from appreciating his roots in the traditional Khasi religion and its practices and ceremonies. To the Khasi society's understanding of their religion authored several years earlier by Jeebon Roy in his *Ka Niam jong Ki Khasi* (1897) and *Ka Kitab Ba Batai Pynshynna Shaphang U Wei U Blei* (1900); Homiwell Lyngdoh added *Ka Niam Khasi* (1937). In this he detailed the ideas of the Khasi religion, its religious performances and the chanting of incantations connected with different ceremonies in the life of the Khasis. A fairly detailed book of 271 pages, *Ka Niam Khasi* also makes reference to the cremation of the *Syiems* of Sohra, the monoliths and the origins of the Syiem and Lyngdoh clans.

Homiwell Lyngdoh followed this marvelous book with his *magnum opus, Ki Syiem Khasi Bad Synteng,* Shillong, (1938). Interestingly he does not refer to the history mentioned above. He has as a reference Joel Gatphoh's *Ka Histori Ki Syiem Synteng Ha Ka Jingshisha* (1923-1926!) which unfortunately is not located. Lyngdoh's book was a trend setter for historical studies in Khasi. The book examined the Khasi-Jaintia political history providing details of each of the Khasi *himas.* He goes into a long introduction of the origin of *Syiemship,* its variations, powers and functions of *Syiems,* the role of the founding clans, the *Bakhraw,* and the linkage of Khasi religion with the

Khasi *himas.* He starts with an account of the founding of the Khasi polity with the establishment of the first of the Khasi *himas,* the *Syiemships* of Madur Maskat, Jaintia and Shillong. The reprinted edition of the book (1964) brings up the introduction to the national movement, the establishment of the federation of Khasi States, the Constituent Assembly of India, the integration of the Khasi States into the Indian Union and the establishment of the District Councils.[11] Then follows in much more detail nineteen chapters in which accounts are given of the 18 *Syiemships.* Each of the shorter histories of the *himas* under *Syiemships* conclude with a genealogy of the *Syiems.*[12]

Homiwell Lyngdoh did not examine the effects of British paramountcy over the Khasi *himas* and their connection if any,with the Federation of Indian States which had emerged in the early part of the 1920s and was strengthened just before the transfer of power from Britain to India. While providing a history of the Khasi *Syiemships* the author does not make much reference to the other Khasi *Himas* under *Lyngsohs,* Sirdars and *Wahadadars.* What is appreciated, however, is that this was the first and elaborate political history of the Khasi and Jaintia states. Its contribution to literature may be assessed from the numerous references other writers have made to this significant publication. A recent review of this book says:[13]

> Though this book continues to serve as a basic text in the study of Khasi- Jaintia political history, its tendency in the retention of the divided nomenclature like Khasi and Synteng did not simply show lack of local exposure especially with the Pnars but displayed far significant colonial influence.

Other Histories

Though Rev. J.J.M. Nichols-Roy, a church elder and politician and perhaps the most prominent Khasi of his times, did not write history, many of his writings with their political bend made use of history. We may here cite just one of his pamphlets: *Ka Ri Khasi Bad Jaintia: Shabar Bad Hapoh Ka Reform.* The lecture was distributed in October 1936 to educate the Khasis of their position in the reformed constitution of 1935. He would continue to write till the 1950s both in Khasi and English on a variety of subjects.[14]

With independence came the freedom to interpret the history of Khasi-Jaintia. In this new approach L.L.D. Basan very early published a bilingual pamphlet *The Khasi States under the Indian Union,* (1948), which questioned the position of the Khasis under the Indian Union and the process by which the Khasi states and British villages had come into an Assam administration. Many years were to pass before L.G. Shullai took up where L.L.D. Basan concludes his account. Shullai campaigned for a relook into the status of the Khasi *Syiems,* the integration of the Khasi Jaintia Hills into the Indian Union and kindred other topics. His *Ki Hima Khasi* (1975) draws much from the references noted above and from the author's perseverance in collecting every scrap of information on the subject of the Khasi *Syiems.* The importance of this small book is that it updates the issue to the present times and has become a ready reference on the Khasi *Syiems.* A variation of this appeared not long after in *Ka Ri Shnong Pdeng Pyrthei* (1989), the introduction of which is addressed to the youth as an inspiration for them to see their society as distinct despite their small numbers. Other writings of L.G. Shullai that have a historical approach are *Ki Symboh History Bad Ka Ri Hynniewtrep* (1989), a collection of essays on aspects of British rule in India; the Assam Legislature 1912-1950, the Constituent Assembly of India, and the administration of the Khasi-Jaintia Hills after 1950 and till 1974.

Rather disappointing as a work of history is I. Nongbri's *Ka History Ka Ri Hynniewtrep* (1992). A close look at the cover clears that the history he writes does not go back to the early past but starts in 1952. This is a repetition of what Shullai and Basan had earlier published; more particularly in the appendices that form a large part of the paperback. To his credit though, are the appendices of interesting letters of Wickliffe, *Syiem* of Nongstoin, who challenged the Government of India's integration of *hima* Nongstoin into the Indian Union in 1947.

Before we again bring the history down to our times it would convenient to go back to the early part of the century and again build up the literature on other aspects of history. With the spread of Christianity in the hills and the establishment of the Seng Khasi (1899) a literature emerged which may be termed revivalist. In this category are the writings of Jeebon Roy in his *Ka Niam Jong Ki Khasi* (1897) and *Ka Kitab Ba Batai Pynshynna Shaphang U Wei U Blei* (1900); Radhon Singh Berry published *Ka Jingsneng Tymmen Shaphang Ka Akor Khasi* (1902) and Rabon Singh Kharsuka's *Ka Kitab Jingphawar* (1905). These stalwarts of Khasi religion and culture are especially noted for they gave the Khasi society a strong influence to remain in the traditional faith.

Just how concerned Homiwell Lyngdoh and his contemporaries were about the past is illustrated in the studies by two Salesian priests. J. Bachhiarello could identify himself with the people sufficiently to author *Ki Dienjat Ki Longshuwa* (1930). Another of his confreres G. Costa endeared himself to the Khasis with his *Ka Riti Jong Ki Laiphew Syiem* (1936). These writings in Khasi have become invaluable texts for scholars wishing to pursue a search into the Khasi past. While at this stage of Khasi religion, it would not be inappropriate to note the histories on Christianity. *Centenary History Ka Balang Presbyterian* was published in 1941. Though many small tracts had been brought out by that time it was only in 1966 that Rev. G. Angell Jones translated and updated J.H. Morris' *History of the Welsh Calvinistic Methodists Foreign Mission* (1910). The Catholic mission history was put in simple language and style by Fr. O. Paviotti in *Ka Matti Jong U* (1989). Fr. H. Elias', *Ka Histori Jong Ka Balang Catholic* (1965) and Francis Diengdoh's *Ka History Jong Ka Balang* (1980) give a general picture of the Catholic Church's history.

It is not intended to exclude the history of Khasi poetry, prose and drama from this account for these genre of literature is as important in the transformation of the Khasis as were other factors such as education, British administration and urbanization. As much as already been done on this by the members of the Khasi Author's Society it may suffice to note that the Khasi literature has grown sufficiently to be reflective of a society's wish that their language and literature receive the recognition of the Sahitya Academy.

Today a trend is clearly emerging in Khasi history of biographical sketches and monographs. Tirot Sing and Kiang Nangbah, two resistance leaders have received a fair amount of adulation and if the present position is any indication, the lives of these patriots will continue to inspire writers.[15] Thomas Jones is beginning to find a more befitting place in Khasi history.[16] J.J.M. Nichols Roy has found his biographer in O.L. Snaitang.[17] Dr. John Roberts a medical missionary is given a sketch by L.G. Shullai.[18] Other Khasis such as Jeebon Roy, Mavis Dunn Lyngdoh already have accounts of them in English and it may be not before long that Khasi readers are made familiar with their lives.

In the absence of comprehensive histories one has to be on the watch for small pamphlets, booklets and paperbacks such as V.B. Ryngnga's *Na Them Riwar* (1992), a collection of 20 short notices in the *War* area. Short histories of the Khasi *himas* are not many. One such is E.W. Chyne's, *Ka History Jong Ka Hima Mawlong* (1977) which is a useful little booklet. Another small publication is Jor Manick Syiem's *Ka Jingiathuh Khana Shaphang Ki Syiem Jong Ka Hima Mylliem 1830-1960* (1984). Spiton Kharakor's, *Ki Khun Ki Ksiew U Hynniew Trep* (1981) makes very interesting reading, particularly the chapters 'Ki Kur U Hynniew Trep' and 'Ki Jait U Hynniew Trep' which enumerates 3363 Khasi clans. Other essays are on a variety of subjects relating to culture, Kiang Nangbah, Tirot Sing and the Khasi National Dorbar.

Conclusion

One reason perhaps why Khasi history has not acquired as much interest as other genre of literature, could be the absence of history in both Khasi and English school texts. An earlier generation of Khasi school goers had to read history texts in Khasi.[19] This made them familiar with the history of India and the world. For young students not familiar with English, this medium enabled them to better understand their own histories. With the change in school curriculum, history has taken a back seat. Consequentially, whatever incentive was there to write history school texts is gone and so has the foundation for an appreciation of the past.

Another point to note is that the histories in the Khasi vernacular may have been written with the intention of portraying the past but many of the authors are not trained in the art of collecting, interpreting and presenting their findings. This has made the better written history rare. Homiwell Lyngdoh's *Ki Syiem Khasi Bad Synteng* was the exception. The book remains a classic for its language, spread of history and the details provided on the Khasi *himas*. Khasi historians, if we may make a categorization, have preferred the English medium for a number of reasons of which the principal would be that English would have a larger number of readers. Before long it is hoped, the Khasis who are becoming increasingly conscious of their roots and culture will find histories relevant to this time and age.

Notes and References

*David R. Syiemlieh, 'Homiwell Lyngdoh (1877-1958)', J. B. Bhattacharjee (ed.), *Historians and Historiography of Northeast India*, Akansha Publishing House, New Delhi, 2012, pp.65-72.

1. For details on the life of Dr. Homiwell Lyngdoh read Maurice G Lyngdoh, *Life and Works of Dr. Homiwell Lyngdoh*, reprinted 2009. A brief biographical sketch is provided in O. L. Snaitang, *A Dictionary of Khasi Literature*, Shillong, 2011, pp.107-111.

2. *Ibid.*, pp.43-47.49.

3. H. Bareh, *The History and Culture of the Khasi People*, revised and enlarged edition, Shillong, 1985, p. 400.

4. Anon, "The Role of the Presbyterian Church in the Development of Khasi Language and Literature 1841-1902', *Khasi Author's Society Souvenir 1842-1992*, pp.12-13.

5. Cited by I.M. Simon, " Tribute to the Rev. Thomas Jones", *Ibid.*, pp. 25-26.

6. D.R. Syiemlieh, " Education, Elite and Politicisation in the Khasi Jaintia Hills", Ashish Bose (ed.), *Tribal Demography and Development in North East India*, New Delhi, 1990, p. 178.

7. D.R. Syiemlieh, *British Administration in Meghalaya: Policy and Pattern*, Heritage Publishers, New Delhi, 1989, pp. 103-106.

8. *Assam Administration Report 1902-1903*, p.44.

9. Cited by L.S.Gassah "Don Katno Tylli Ki Elaka ha Ri Jaintia", *Hynniew Trep Endeavour Society*, September 1989, p. 33; Suniti Kumar Chaterjee, *Kirata Jana Krti*, Asiatic Society of Bengal, Calcutta, 1974, p. 172.

10. See a related article on the Sohra *Syiemship* in D. R. Syiemlieh, "Colonialism and Syiemship Succession: a Study of Cherra State (1901-1902)," *Proceedings of the North East India History Association*,Third session, Imphal, pp.147-157.

11. *Ki Syiem Khasi Bad Synteng*, "Jinglamphrang," pp. i-xvii and "Ka Juk Thymmai," pp.i-vii.

12. Homiwell Lyngdoh must have drawn heavily from D.Herbert's *Report on Succession to Siemships in the Khasi Hills* (1902) which oddly is not listed in the references of book used to write this history.

13. O.L. Snaitang, op. cit., p 110.

14. For a collection of the writings of J. J. M. Nichols-Roy, read O. L. Snaitang, *Memoirs of the Life and Political Writings of the Hon'ble J. J. M. Nichols- Roy,* Shillong, 1998.

15. Hamlet Bareh, "Ka Thma Synteng Bad U Kiang Nongbah,"(1961); Hamlet Bareh, *U Syiem Tirot Singh bad Kiwei Pad Ki Para Syiem Ki Hima Khasi* (1977); Hamlet Bareh, *Shishpah Sanphew Snem Mynshwa: U Tirot Singh bad Ka Mynnor ba Iphuh Ipheng* (1984); Quotient Sumer, *U Kiang Nongbah* (1978) are only several of the many articles that have been published in Khasi in souvenirs/tracts remembering these two Khasi and Jaintia leaders who struggled against the British. The literature in English on this subject is large and better researched.

16. Refer to the many articles in the *Khasi Author's Society Souvenir 1842-1992* ,and S.S. Majaw, *U Kpa Ka Thoh Ka Tar Khasi*, Shillong,1992.More recently was published S. S. Majaw, *U Thomas Jones Bad Ka Pyrthei Saitpohpen*, Khasi Book Stall, 2011.

17. O. L. Snaitang, *Ka Biography U J.J. M. Nichols-Roy*, Shillong, 1993.

18. L.G. Shullai, *U Dr. John Roberts DD,* Shillong, 1975.

19. Prescribed for schools were Frank Pugh, *Ka Histori Jong Ka Khasi Jaintia Bad Assam* , Shillong (n.d.) and Tngensi Rynjah, *Ka History Ka Ri Khasi Jaintia*, Shillong, 1991.

14

Indigenous Roots: Hajom Kissor Singh and the Founding of Unitarianism in the Khasi-Jaintia Hills

Introduction

Like that of many religious denominations, the founding of Unitarianism in the Khasi-Jaintia Hills centres on the role of an individual-*Babu* Hajom Kissor Singh. He established a church at a time when the Khasi-Jaintia society was on the cross-road towards what was the first and, for some time the only Unitarian mission in South Asia[1], different in some of its beliefs with that of the main line Christian churches yet liberal[2] in approach and worship. The advent of Unitarianism in the Khasi-Jaintia was a low-tone breakaway from the Welsh Calvinistic Methodist (Presbyterian) Church. To emphasize the role of an individual would be limiting in a study of the founding of Unitarianism in these hills, for its strength came initially from a small band of followers growing in time to be recognized as one of the important faiths of a section of the Khasi-Jaintias. The foundation therefore should be examined in the broader context of the religious, cultural, social and intellectual ferment that the Khasi-Jaintia community was experiencing in the last quarter of the 19th century and in the early years of the 20th century.

Society in Transition

The two quarter centuries bridging the 19th and 20th centuries was a remarkable period for the Khasi-Jaintias. It marked the development of Khasi literature which

Welsh Presbyterian missionaries had first put into Roman letters, encouraged in their translation of the Bible and other Christian literature and flowering into prose, poetry, drama and song by missionaries and their mission school products.[3] Much of the emerging literature had a strong base in the religion and culture of the society. Christian and the adherents of Niam Khasi were together involved in discussing the changes and response the society was undergoing. The newspapers that reflected their mood was *U Khasi Mynta*, published by Jeebon Roy from Ri Khasi Press. Started in August 1896, this paper and *U Nongphira* which was first circulated in 1903, kept up a relentless crusade to preserve Khasi culture and religion.[4]

Linked with the growth of the Welsh Presbyterian Mission was the spectacular growth of mission schools. Wherever a mission was established a school was invariably started. Before the turn of the century mission schools were operating across the entire Khasi-Jaintia Hills. From 290 day scholars in 1861 the number of students in Mission schools reached 4626 in 1891.[5] By 1900-01 there were 325 schools with 6535 students in the district.[6]

The nineteenth century, Indian social and religious reform movements reached as far east as Shillong and Cherrapunji. The movement came by way of Bengalis attached to the Assam provincial administration. Many among this community were Brahmos. They had set up residence in and around Laban, organised places of meeting and worship, and entered into dialogue with the more educated Khasis. Occasionally these Brahmos would be visited by their own missionaries. Babu Nilmoni Chakravorty was sent out by their mission and resided in these hills for a number of years. The Khasi religious and social change was a combination of many factors. It saw expression in the founding of the Unitarian Church in 1897, which will be taken up presently, and the Seng Khasi in November 1899 aimed at fostering "brotherhood among Khasi who still retain their social-cultural and religious heritage."

Influences and Foundation

Hajom Kissor Singh was a product of this era. He was witness to the changes his society was undergoing with all its influences. Born on 15 July 1865 in Saitsohpen, Cherrapunji, he matriculated from the mission school. At the age of 15 he became a convert to Christianity. Hajom's father, Bor Singh was employed as police Sergeant at Jowai. Young Hajom lived in Jowai after his schooling. Before long he too found employment in Jowai as *Amin* in the sub-division office. Within a year he was promoted as Clerk-cum-Surveyor. Another promotion came his way in 1888. Between 1891-1904, he served as Head Clerk in the same office. The Government took notice of his work when in 1902 he was made *Dewan* of *Hima* Khyrim,[7] a rare distinction for a Khasi at that period of time.

Hajom was well read, inquisitive and extremely hard working. A cursory glance into his diaries covering his service in Government and Church indicate his love for books that he would order from Calcutta. His reading ranged from philosophy, history, geography and religion. He was a practicing dispenser of medicines and a keen observer of the natural phenomena around him. Some of his diaries go into long accounts of his visits to village in the Khasi-Jaintia Hills; there is an especially long narration of a visit he once made to Jaintiapur, the former capital of the Jaintia state, in Sylhet. His note on the earthquake of 12 June 1897 unfortunately has not been noticed by historians, geographers and geologists who have written on that most disastrous quake.

Hajom's disenchantment with the Welsh Calvinistic Christianity must not have come very suddenly. There is no account of how this was brought about other than his contact with a Khasi Brahmo, Jope Solomon who directed him to get in touch with Rev. Charles Dall, an American Unitarian Missionary at Calcutta. Even before contact was made with the missionary, Hajom had questioned certain beliefs of the Christianity he had accepted.[8] Rev. Dall sent him some literature on Unitarianism including a volume of Williams Channings writings on the subject. "As a consequence of his study of the literature sent to him, his faith in Unitarianism became confirmed and his understanding of it enlarged. He became a missionary of his new faith to others,"[9] The beginnings were small. On 18 September 1887 at Jowai, Hajom formed a community of believers, a nucleus with three members, Herbon Lakadong, Mar Sutnga and Kat Shilla.[10] An interesting letter Hajom Kissor Singh wrote in 1916 has been overlooked by Unitarians tracing their heritage. Giving an account of the 29th annual conference of the Khasi Hills Unitarian Union held at Jowai. 5-7 February 1916, he wrote:[11]

> I find however that the date 10 September was entered against the name of my wife I Pharien Bon in the first Roll Register of the Union. This shows that we had made a vow to be Unitarian before we could hold any Unitarian meeting. I still remember the discussions of those early days of the Union and the question she used to ask me about the deity of Jesus.

This letter establishes that the Unitarian church was started on 10 September 1887 and that Hajom Kissor Singh and his wife were its first adherents.

The founding of Unitarianism in the Khasi-Jaintia Hills "was the result of similar Unitarian ideas developing separately in the minds of two men," wrote Griffith Sparham.[12] The other was Heh Pohlang, a member of the Presbyterian Church at Nongtalang. He too had been reading and questioning his faith. It was somewhat fortuitous that the two men should meet and the Church expand into Nongtalang - the largest Jaintia village in the southern face of the hills, overlooking Jaintiapur.[13] By 1890, there were 11 numbers in the Nongtalang Church. Another five members made up its strength in Jowai.[14] Some years earlier Shillong house was formally opened in Sunday, 17 July 1898, when about 100 people attended.[15] Statistics of the Union till May 1899 show that the congregation at Jowai, Nongtalang, Rilang, Laitlyngkot, Shillong, and other villages numbered 204 believers with 4 preachers.[16]

Unitarian was making itself felt in these hills not without some opposition from the dominant church and "with no western resident to aid, counsel, or control."[17] Unitarianism in Shillong preceded Hajom Kissor Singh's transfer to the Deputy Commission's Office and his residence in Upper Laban in 1904. One of *Babu* Hajom's converts Robin Roy established a regular congregation in the town moving to Madan Laban in 1896. The entire congregation in Laitlyngkot, composed of tobacco traders moved to Mawpat in 1903 where they erected a small church. The headquarters of the Khasi Hills Unitarian Union was shifted to Shillong in October 1906.[18]

Indigenous Roots

That Unitarianism had indigenous roots is made clear by the founder, his American supporters and historians. Hajom Kissor Singh wrote in his diary of 1906 that the Unitarian Union was founded "without' any foreign aid."[19] Rev. J.T. Sunderland noted in one of his accounts of the start of the Church in these parts:[20]

> Nothing has arisen in the history of Unitarianism for many years more interesting than the Unitarian Movement that has sprung up within the past half decade as a purely native growth on the soil of Assam, India.

Nalini Natarajan starts the section of her book 'Locally Born Christian Sects: Unitarian' that "The Unitarian Movement in the Khasi Hills was entirely locally born"[21] Two scholar who have researched on the subject have given details of its indigenous beginnings.[22]

Could this fledgling church have continued without support from foreign missions? From a reading of the growth of the church into the 20th century it becomes very clear that the Unitarian Church in the Khasi-Jaintia Hills at all times have managed their own affairs. There is no doubt that an early stimulus was provided by the American Unitarians and that visits from co-religionists encouraged their endeavors. Reverend Sunderland was in correspondence with Hajom Kissor Singh. He often made references in the journal he edited of the efforts of the Khasi Unitarians.[23] Rev. Sunderland visited Jowai in 1896. His visit formally established the Unitarian Union. Rev. James Harwood of England visited Jowai in 1897. Two years later another English Unitarian Missionary Rev. S. Fletcher contacted the believers and encouraged them to establish an Endowment Fund for the Union. In November 1893 the support of the Khasi Hills Unitarian Mission was transferred by the American Unitarian Mission to the British Foreign Unitarian Mission.[24] From 1894 the British Unitarian Association sent aid to the Khasi Hill Unitarian Association. The aid was to carry on native mission work in the hills. When the British Foreign Unitarian Association began to interfere in the internal affairs of the Union, the Board of Directors meeting in Jowai in March 1904 expressed its disapproval of the interference; and at its Laban meeting held on 9 July 1905 declined the conditional aid of the B.F.U.A. "Preferring to be faithful to the principles of the Union than to accept a Non-Unitarian as our Ruler.[25] The issue was the support by the British Foreign Unitarian Association of Nilmoni Chakravarty a Bengali Brahmo, as Superintendent of the Unitarian Union, a decision the believers could not accept.

Funds collected from the believers were small. The churches infrastructure was largely supported by the congregation composed of very simple and ordinary Khasi-JaintiasTo give just an instance of its financial straits; the church needed a hymn book. Rs. 19.5.6 was collected from the believers. Rs. 71.14 was collected from the Americans Union through the efforts of Rev. Sunderland. 500 copies of the hymnal were printed. When it arrived and put to used "it was informally agreed to omit the verse regarding Jesus when repeating our principles of faith as contained in our Hymn No.1," writes the founder.[26] He diarized that Babu Durga Singh was also very against repeating the name of Jesus." I have observed the change in my friends' mind for some time. Of course there will be no loss in omitting the name of Jesus because his teachings and spirit must be retained," he notes.[27]

Hajom Kissor Singh retired from Government Service in1922. For many years he was the Union Treasurer. He passed away on 13 November 1923 while on a tour to Puriang. He was cremated to enable his ashes to be brought to Shillong.

Conclusion

The Unitarian Church in the Khasi-Jaintia Hills is not as large in numbers as some other congregations and fellowships started in the same hills many years after Unitarianism was founded. Perhaps one could say it is not numbers that matters but the reasons and the roots, the spread of the Unitarians, its religious, social and economic impact and the test of time as testimony of a religious persuasion which has very much become part of the life of these hills.

Notes and References

*David R. Syiemlieh, 'Indigenous Roots: Hajom Kissor Singh and the foundation of Unitarianism in the Khasi-Jaintia Hills', O. L. Snaitang (ed.), *Churches of Indigenous Origins in Northeast India*, ISPCK, Delhi, 2000, pp.150-157.

1. Nalani Natarajan, *The Missionary Among the Khasis*, Sterling Publishers, New Delhi, p.81.

2. *The Unitarian*, a magazine of "Liberal" Christianity was one of the important supports base for the Khasi-Jaintia Unitarians.

3. Among the more prominent contributors to Khasi literature were William Pryse, William Lewis, John Robert, all missionaries, and Radhon Singh Berry, Jeebon Roy, and Soso Tham among the more prominent Khasis of their time.

4. See J.B.Bhattacharjee, 'Socio and Religious Reform Movements in Nineteenth and Twentieth Centuries: Khasi and Jaintia Hills', S. P. Sen (ed.), *Social and Religious Reform Movement in the 19th and 29th Century*, Institute of Historical Studies, Calcutta, 1979.

5. John Hughes Morris, *The History of Welsh Calvinistic Methodists' Foreign Mission to the end of the year 1904*, Indus Publishing Co., New Delhi,1996, p.192. Some schools had been started and maintained by Government of which the Normal School at Cherrapunji and the Government High School at Mawkhar, Shillong, started in 1878 were prominent.

6. B.C.Allen, *Assam District Gazetteer*, Vol. X; Shillong, 1906, p.109.

7. Griffith Sparham, *Khasi Calls,* The Lindsay Press, London, 1945, p.11; Renewlet Nongbri, *Growth and Development of Unitarianism in the Khasi and Jaintia Hills*, Shillong, 1989,p.7.

8. R.Nongbri, op.cit., p.8.

9. *The Unitarian: A Monthly Magazine of Liberal Christianity*, Editor, I T. Sunderland, vol.vii, Boston, 1892,p.369.

10. R. Nongbri, op.cit., p.10.

11. *The Christian Observer*, 22 June 1916, A gist of the letter was also published in *The Christian Life*, 10 June 1916.

12. Griffith Sparham, op.cit., p.3.

13. *Ibid.*, p.4.

14. *The Unitarian*, Vol. VI, Boston, 1891, p.I72.

15. Hajom Kissor Singh, Diary 1897-98.

16. Records with Kong Imelda Ranee, granddaughter of Hajom Kissor Singh, in a file relating to Census Reports; also Griffith Sparham, op.cit., p.7.

17. Griffith Sparham, op.cit., p.7.

18. H.K.Singh, Diary, 1906.

19. *Ibid.*, p.58.

20. *The Unitarian*, Vol.VII, 1892, p.l77.

21. Nalini Natarajan, op.cit., p.81.

22. R.Nongbri, op.cit. The text of this book was first submitted as an M Phil dissertation to Department of History, North Eastern Hill University; Danibha, Pyrbot, 'Unitarianism in the Khasi-Jaintia Hills 1887-1995',Phd NEHU 2012, pp.45-50,69.

23. *The Unitarian*. Vol.VII,1892, p.369, made mention of the Nongtalang Church and the appointment of Riang Pohlong as lay preacher.

24. *The Christian Register*, 22 June 1916.

25. H. K.Singh, Diary 1906, pp.59,61; Diary 1904, p.20

26. H. K. Singh, Diary 1904-1905, p.97. *U* Kyllang Padu was appointed President of the Union at the Nongtalang Conference, 1 January 1905.

27. H. K. Singh, Diary 1906, p.87.

15

David Roy:
Notes on the Khasis

Introduction

The Khasis came under British imperial rule in the early part of the nineteenth century. Their hills were not formally annexed into the colonial state. The Khasi *himas* remained as subordinate and dependent states under the paramount power. Twenty-five *himas* were placed in this position of indirect administration. Their *Syiems, Lyngdohs, Sirdars* and *Wahadadars* administered their states in their own genius but under the watchful eye of the colonial administration. The Jaintia *hima* was annexed and made part of British India in 1835. In time several villages were taken away from their *himas* and made into British villages and administered by *Sirdars* directly under the district Deputy Commissioner.

Other than the Khasi resistance to the expanding British position in 1829 -1833 and the Jaintia rebellion of 1860-1862, there was no serious opposition to colonial rule in these hills. The Khasi-Jaintia quietly acquiesced themselves to British rule. Special provisions were made under that administration for ruling the hill people of the region including the Khasis-Jaintias, which in part explains the orderly administration for the duration of British rule. With little investment in terms of funds and personnel, the British could turn the tribes into passive servants of the *raj*. The spread of Christianity in these hills must have added to the acceptance of foreign rule. Orderly administrations, the operation of the rule of law, one economy under one political system apparently were the hallmarks of the British *raj* and its administration.

Independence placed the Khasi-Jaintia Hills in the state of Assam, much to the concern of the leaders of the tribe. They had wanted to remain outside Assam but legislators and policy and decision makers after 1947 had other ideas. They positioned the hill people under Assam. This too the Khasi-Jaintia accepted without much opposition, which even if there was concern was but feeble, rhetoric and petered out very soon. Political consciousness, however, was taking roots. It had started in the 1920s with the Government of India Act 1919 entering the statute book. A new form of politics entered these hills- electoral politics. This was to be a new and different form of governance the full significance of which was to come after independence and the elections of 1952. A new class of leaders emerged principal of who was J.J.M. Nichols-Roy. He and others including several British officers and David Roy focused attention on the future of the tribe and raised issues relating to a variety of questions on their governance. J.J.M. Nichols-Roy was heard because of his position as a clergyman and political affiliation. The British views were not considered. Their plans for the region were shelved and little discussed by the new Indian leadership.[1] Three of David Roy's papers on these issues, which will be discussed at some length, appeared to have had little effect other than expressing a tribesman's concern.

The connection of the Khasis and Assam was but brief. The Assamese leaders' policy towards them becomes very clear from the debates at the Constituent Assembly. One problem with Indian democracy in the early phase of its operation to take a case, is that it enabled the dominant Assamese community to pursue policies detrimental to the unity of the state. The breakaway was only a matter of time before the hill people decided their destiny would not be with Assam.

When the Hill State movement was well under way in the late 1960s, the Government of Assam had to recognize that it was a legitimate demand of the hill people to ask for a state of their own. The question was then asked by several leaders what should be the name of the autonomous state soon to emerge? Very appropriately the name 'Meghalaya' was chosen. Its author was S.P. Chatterjee, a geographer who came to these parts in the late 1920s. He called the plateau where the Garos, Khasi and Pnars reside "Meghalaya," the adobe of clouds. First used in a doctoral dissertation submitted to the University of Paris, Sorbonne, in 1939,[2] the term has come to denote the state and its people. Unfortunately not much has been researched on who first made this suggestion, and the discussions in the Central and Assam legislatures and whether it was discussed among the tribal communities who would come under this nomenclature.

Nurturing the Tribal Intellect in David Roy

David Roy was born on 23 December 1884 in Laban, Shillong. The son of Rai Bahadur Roy Singh (Shabong), David Roy passed the matriculation examination of Calcutta University in March 1901 with a first division. The young Roy must have been witness to the exciting but not too tumultuous times that Khasis were undergoing with British rule firmly established; the spread of Christianity in his hills and the concerns of the followers of the Khasi faith such impact was having on the people. He would have been the second generation of Khasis, and there would not have been many, to have had the benefits of reading Khasi in print. Roy spent a period of time in Calcutta pursuing his studies. It appears he did not graduate as he was offered and joined the Assam Civil Service in 1904. Service took him to the Naga Hills, Garo Hills and the then Lushai Hills districts. He was for some time Sub-division Officer at Jowai. He retired from government service as Extra Assistant Commissioner in 1941.Early that same year he was conferred the distinction of Master of the British Empire, a rare honour for an Indian and Khasi.

His administrative acumen must have been noticed because no sooner had he retired from service, Government appointed him *Dewan* of the then called Cherrapunji state. The following year he was given additional charge as *Dewan* of Mylliem *hima*, continuing in this position in the latter till 1946 and as *Dewan* of Sohra till the year of independence. The only other Khasi to have held such an office was Hajom Kissor Singh, the founder of the Unitarian Church in the Khasi Jaintia Hills, who served for a period of time as *Dewan* of *hima* Khyrim.It was noted in an official correspondence signed by the Governor of Assam that David Roy brought in good governance to both Sohra and Mylliem, improving the finances of the *himas* before the administration of these two Khasi states were returned to the traditional chiefs. A devout Christian still attached to his cultural roots, David Roy passed away on 13 February 1966. He was buried in the Anglican Church cemetery in the Shillong cantonment.

With this background we may focus on David Roy and his views on the position of the Khasis shortly before and after 1947. An eloquent speaker and an erudite writer, David Roy was among the first of Khasis to have found a place as an administrator under the Raj. There were several other Khasis who rose in rank under the administration. *Babu* Jeebon Roy was the first. Much has been written on him and his achievements. However, little is known of David Roy. One reason for this was that David Roy did not compile his writings in a volume or two. The articles he authored are to be found in several academic journals and no attempt has been made to collate these. His grandson, Malcolm David Roy has recently published a brief biography of the man and reprinted the articles he authored.[3] Perhaps little is known of the man because for much of his service he was serving in other parts of Assam than in his own hills. By far though, we know so little of the man because he did not have use of any political platform to voice his concerns and views. Were he to have addressed audiences as did his contemporaries he would surely have been recalled.

This essay is an attempt to fit the man in his own time and situation and refer to his writings in a bid to give him a place among the first of Khasi intellectuals. As with many other young men of his time, David Roy was nurtured in his family and clan, his faith and his environment in these hills and his administrative service. These influences in his early and adult life are reflected strongly in his writings. We will focus largely on the ideas of the man on the future of his hills in an independent and free India. This essay is a tribute to a Khasi long forgotten.

Span of Mind

It would be in order to say something of David Roy's writings on Khasi religion and society before discussion turns to other ideas. After P. R. T. Gurdon, Jeebon Roy and Sib Charan Roy it was David Roy who wrote extensively on Khasi religion and culture. Roy wrote largely in English. If a chronology of his writings is taken then his first was 'Principles of Khasi Custom', a seminal piece in Keith Cantlie's *Note on Khasi Law.* [4] This has endeared him to his people for it was among the first discussions on the topic of Khasi customs written by a Khasi. In this he discussed the results of contacts with outside influences; the position of Khasi women; the Khasi idea of marriage; the Khasi interpretation of the human race; the position of man in the family; the Khasi idea of life and his *niam* and distribution of property, among others. In December 1936 was published 'Principles of Khasi Culture.'[5] It provides an overview of Khasi culture and should be essential reading for anyone interested in this subject. Two years later came another long article 'The Place of the Khasi in the World.'[6] This confirms the deep sense of respect and authority he had of Khasi life and customs. A brilliant piece on Khasi megaliths was published in *Arthropos*. As an essay on megalithic culture seen through the eyes of the author, "the article is invaluable" wrote C. Von Furer-Haimendorf [7] This essay more than any other established David Roy as an authority on Khasi culture, something that has not been appreciated sufficiently. Recognition of his scholastic ability came in his time. Among the several honours he received was a Fellow of the Royal Anthropological Institute of Great Britain and Wales in 1936.

Khasi Hills 1946-1947

By 1945 it had become clear that the British were to leave India. There was hectic political activity in the Khasi and Jaintia Hills as elsewhere, before the transfer of power. McDonald Kharkongor started a Hills Union, which, however, never became a sustained movement. The Khasi-Jaintia Political Association with the veteran leader Dr. Homiwell Lyngdoh as its Secretary took the cause of the people at large and particularly directed its attention against J. J. M. Nichols-Roy's Khasi Jaintia Federated States National Conference. The Khasi-Jaintia Political Association looked forward towards independence but did not want the change to result in others ruling over their hills. The Khasis expected much from Jawaharlal Nehru's visit to Shillong early in January 1946. His refusal to acknowledge any claims of separateness or even to consider the existing restrictions on acquisition of land by plainsmen dampened the Khasi-Jaintia enthusiasm and set back by many years the Congress interest in these hills. At one of the meetings Nehru is said to have left without addressing the gathering.

The Indian States in general, were once again activated when the Cabinet Mission submitted its Memorandum on States' Treaties and Paramountcy of 22 May 1946.The Memorandum stated that with the transfer of power the British crown would cease to exercise paramountcy and that the rights of the Indian states would return to them. The void that would arise from the lapse of political arrangements between the States and the Crown would be filled in either with the States entering into federal relations with the succeeding Government of Governments or enter into political arrangements with or without them. The States were free to associate with one or the other of the Dominions of India or Pakistan, to federate among themselves or stand alone.

In this background the Federation of Khasi State first set up in 1934 was revived on 22 August 1947. This development followed a meeting of the Khasi *Syiems* held in Shillong on 22 August 1946 at which it was decided to have a Federation of Khasi States to enable them to have external relations, promote good Government in the States and safeguard their interest. All these parties agreed that the Khasi States should be brought into some sort of federation affiliated to the province of Assam.

David Roy by now retired but continuing as *Dewan* of Sohra and Mylliem *himas* was a witness to these developments. In November 1946 he published a small tract entitled *Whither the Khasi Hills?A Study.* [8] What is of interest to us is the second part in which the author goes beyond the title of the tract to ask "and where is the leadership?" He asked another question: "can they (the Khasi States) still exist as small states and solve their problems?" David Roy does not make any reference to the possibility of the collapse of imperial rule. He could not mention this presumably because of his position and past service. Roy noted the problem of several small *himas* "each independent of the other." He called for unity of the twenty- five *himas* and the British villages. Suggestions were then being discussed in political circles of a federation. To his mind he wanted the federation to link not only the *himas* but to also include the British villages. The basis of the proposed federation to his mind was to set up industries. " Khasis will need industries and industries will need mechanics and mechanics will need energy," he wrote. He called for the harnessing of hydro-power and other resources with a mention of coal and limestone. Could all this be done by private enterprise and whether the states and British areas were large enough in finance and capital, he asked. In his scheme of things Roy wanted Khasi culture to be the basis and common unit of the federation.[9] Moved by the argument he builds up, Roy wrote:[10]

> It is not a mere abstract question of advantages and disadvantages of isolation that we are debating. It is the vital question of survival. It must go forward and not stand still. The Khasi nationalist who regards his people as a separate race with a culture of its own need not fear the taunts of those who call them a foolish friend of the anthropologists who wants to keep us as a primitive people for a museum or laboratory. Khasis are too forward in their ideas of democracy and autocracy to need a copy from anywhere.

Just shortly before the transfer of power, David Roy published a more detailed tract, *The Frontiers of North-East India.*[11] Some of his ideas mentioned in the first two pages of this tract emphasised the issues raised in the tract published in the year before. He then goes into a discussion on the future of the tribes of the North East in India. He gave a call to treat the North East as one group and not as disconnected entities.[12] The crux of his argument and concern is in this line: "The necessity of the solution by some unification is that the tribal areas, though vast in extent, are only vast tracts of no man's land compared with the great continent of India with no more land for man." His fear was that these no man's tracts will be swallowed up by man from the country of "no land for man." He saw a second reason for unification. The tribal areas had neither the man power nor wealth or the means of exchange in the economic sphere with people east of the North East frontier or with their western neighbours. He makes a passionate plea that the law of the land keep even the most weak parts and bring that part to its active consciousness while giving all its constituent part widest latitude for self expression and freedom. Writing at the time when the Constituent Assembly was meeting, Roy urged that the constitution should take note of the special nature of the region and "to assimilate and not to disintegrate."[13]

Roy urged that the solution in the Khasi Hills was in the treatment of the Khasis in the British areas and the Khasi States as one people by abandoning the division of

British and non-British and the treatment of all Khasis to include the Jaintias as one unit. He could foresee that if there was not a uniform system of administration there would be difficulties in the administration of the police and law and order.[14]

As Independence drew near, in early July 1947 agreements was reached between Sir Akbar Hydari the Governor of Assam and the twenty-five Khasi states on the necessity of maintaining the unity of India by the states joining the Indian Union for defense, foreign affairs and communication. A few days before Independence they signed the Standstill Agreement. The Chiefs agreed that from 15 August 1947 all existing arrangement between the province of Assam and the Indian Dominion on the one hand, and the Khasi states on the other should continue to be in force for a period of two years or until new or modified arrangements would be arrived at. Getting the states to sign the Instrument of Accession proved a difficult job for the Governor. Twenty chiefs signed the Instrument of Accession at his official residence on 15 December 1947, and some force and tact got the remaining five to do the same in the early months of 1948. Generally, the Khasi states had no desire to join Pakistan. The *Syiem* of Cherra it is reported did flirt with the local authorities in Sylhet before signing the Instrument of Accession but was warned by Hydari against playing the game. Many of the Khasi chiefs felt that their states had much to lose with the Radcliffe Award, because the new international boundary with East Pakistan along Sylhet and Mymensingh districts was based on geographical and not ethnic considerations. The Khasi states abutting on Sylhet had much of their territory along the foothills and into the Sylhet plains. Their *hats* were located along what would become the boundary. The *Syiem* of Nongstoin's resistance and eventual acceptance of the Instrument of Accession, however did not stop his nephew from making an appeal to the United Nations Organisation, concerning the way in which Nongstoin was intimidated into joining India. Pakistan meanwhile had embarked on a virtual economic blockade of the Khasi and Jaintia Hills. The producers or oranges, pineapples and *tezpat* found no market for their produce. The import of rice was stopped. Such trade restrictions had an adverse effect on the economy as most of the produce were perishable items. The object of the blockade was to put pressure on the hillmen and create among them a feeling that they could be better off in Pakistan. The Indian Government immediately came to the rescue of the producers by constructing an airstrip, which enabled the produce to be airlifted to Calcutta, and by dispatching much needed rice.[15]

The Khasi States had acceded to India but could take no decision to merge their States. Sardar Vallabhbhai Patel visited Shillong between 1 - 2 January 1948. His meeting with the chiefs ended in a stalemate over the merger issue, for they said that only a duly constituted *dorbar* of the States could decide on such a move. Sixteen months later the Khasi States Constitution Making Durbar was inaugurated on the 29 April 1949.It was left the difficult task of arranging an amicable solution to the dual problem of integration and merger of the Khasi *Himas* into the Indian Union. In reality the situation was such that Delhi did not give serious thought to the position of these hills and arranged the integration by administrative and legislative powers of the new Government.

Last Lines

In the evening of his life David Roy still courteous and polite, with never a hard word, reflected on the position of his hills in the Union of India. In a delightful article entitled 'Ka Khyrdop', (The Gate), he muses on Tyrna village located to the south of Sohra and the Shella Confederacy. We find a deep sense of peace in the man and his philosophy of the role of religion in one's life. He beautifully links religion with the political position small tribes such as his own were facing twelve years after the promulgation of the Constitution. He refers to the Union of India, the summit of all the different communities in India and questions as he often does in his essays, "whether such a State would not give rise to destruction in some respect of a part or whole of some one's culture." To quote him in some length:[16]

> A wise administration of minority peoples with a some-what dissimilar culture thrown in the midst of more numerous numbers of a union is an anxious care on the part of the latter to avoid imposing on the minor people any measure which might raise a conflict in the mode of life or culture of such a people, which ultimately will be harmful to their well being.

He continued that "there should therefore not be any form of religious imposition on the weaker section of the community."

Looking Ahead

It may be of interest here to refer to the note of J.P. Mills, Secretary to the Assam Governor. Writing in 1945 he presented three possible alternatives for the future of the hills districts of Assam. First, the inclusion of all the hills in Assam which to his mind was clearly impossible. Under the second alternative, the choice appeared to lie between "the two extremes of excluding only some of the present tribal areas and to including only the present partially excluded areas." He expressed a hope that all the hill tribes would eventually join Assam and be united under one Government and "it may be argued that to postpone the union of any now is to make the ultimate union of all more difficult." He favoured that the union of all the hills be done at the same and appropriate time. The third alternative he discussed was that of temporarily excluding all the hills from Assam and according them special treatment designed on the lines of indirect rule to develop those indigenous institutions which still survived and to fit the tribes for eventual union with the province of Assam.[17]

If the hills were to be separated from the plains, Mills continued, the two alternatives appeared to form a province composed of Districts, Frontier tracts, Agencies and States or to form a union of States from the outset. He strongly favoured the second alternative. Giving an outline of the scheme he desired, he suggested that homogenous areas could be termed " States" such as an Abor State covering all the tribes included under the term; a Lushai State, a Khasi State which interestingly could include the Garos and the Jaintias. Within this union of States would be Sub–states, a term borrowed from the Burma context where small or intermingled tribes had been combined in a single Sub-state. Full use would be made of existing customs and institutions such as the village councils by keeping power in ordinary affairs as local as possible. There would be a gradation of levels of administration from this level to the Sub-state Council, the State Council and the Governor's Council. Mills preferred the use of the term Union of States rather than Province because it was likely to facilitate the inclusion of existing States which in a province would be reduced to the status of districts or not be permitted to continue in the ordinary sense. Moreover, the term "State" might engender a feeling of responsibility and cohesion of the people. With regards to staff each state would require a Political Officer with an additional Political Officer in the larger states. Two officers with the status of Commissioners for the charges North and South of the Brahmaputra would oversee the administration. The object of the administration being to guide rather than to govern, Mills wanted the administrative staff to be as small as possible. In line with the principle of paternal administration the aim ought to be indirect rule rather than administration. For several reasons Mills ruled out Shillong as the possible capital of the Union of States including the possibility of Shillong remaining the capital of the Province of Assam and the difficulty of having an airstrip nearby. Imphal appeared to him to offer better advantages. From there the States could be reached by air, Tripura and Sikkim if included could be reached with equal ease and the Naga State would be within close reach.[18]

Mills with some foresight had suggested the formation of small states. His 'Abor State' has in time become the larger Arunachal Pradesh. Likewise the suggestion he gave of a 'Khasi State' to include the Garos and Jaintias has taken shape in Meghalaya. So too has the 'Lushai State' become Mizoram and the 'Naga State' has become Nagaland.

Conclusion

The history of Meghalaya is yet to be written. When it does take shape it will require recalling the contribution of many in its making, not the least by David Roy.

Notes and References

*Presented at a seminar "Sixty Years of North East-Revisiting the Experience" organized by Divya Jeevan Foundation Aurobindo Institute of Indian Culture, Shillong, 12-14 November 2009, and published in Malcolm David Roy, *David Roy: A Khasi Remembered*, Don Bosco Press, Shillong, 2012, pp.273-286.

1.   For more details of the Crown colony /Protectorate plans read David R. Syiemlieh, *On the Edge of Empire: Four British Plans for North East India1941-1947* , Sage India Publications, New Delhi, 2014.

2.   The thesis was titled 'Le Plateaux de Meghalaya.'

3.   Malcolm David Roy, *David Roy A Khasi Remembered*, Don Bosco Press, Shillong, 2012.

4.   *Note on Khasi Law,* Shillong,1931, reprinted Shillong n.d. pp.84-99.

5.   'Principles of Khasi Culture,' *Folklore*, vol. 47, No. 4., pp. 375-393.

6.   'The Place of the Khasi in the World,' *Man in India*, Vol. 18, Nos. 2-3, 1938, pp. 122-134.

7.   'The Megalithic Culture of the Khasis,' *Anthopos: International Review of Anthropology and Linguistics*, Vol. 58, 1963, pp. 520-556. The article has been reprinted in the *NEHU Journal,* Vol. VIII, no.1, January 2010. pp. 1-13, and Vol. VIII, no. 2, July 2010, pp. 1-20.

8. David Roy, *Whither the Khasi Hills?A Study,* Ri Khasi Press, Shillong, 1946.

9. *Ibid.*, pp. 5-6.

10. *Ibid.*, p.7.

11. David Roy, *The Frontiers of North-East India*, Ri Khasi Press, Shillong, 1947.

12. *Ibid.*, p.4.

13. *Ibid.*, p.5.

14. *Ibid.*, p. 6.

15. David R. Syiemlieh *British Administration in Meghalaya: Policy and Pattern*, Heritage Publishers, New Delhi, 1989, pp. 194-205; Helen Giri, *The Khasis Under British Rule 1824-1947*, Regency Publications, New Delhi,1998, pp.224-262.

16. Malcolm David Roy, op. cit., pp.198-199.

17. J.P.Mills, *A Note on the Future of the Hill Tribes of Assam and the Adjoining Hills in a Self-Governing India*, (September 1945), in David R. Syiemlieh (ed.), *On the Edge of Empire: Four British Plans for North East India 1941-1947*, Sage Publications, New Delhi, 2014, pp.106-115.

18. *Ibid.*, pp.115-134

16

Traditional Institutions of Governance in the Hills of North East India: The Khasi Experience

Introduction

Over the past several years traditional institution of governance in the region and more particularly in Meghalaya have become active. They are finding space in modern governance. The Central government is giving them attention with funds for infrastructure and development. The State administrative machinery depends on the institutions for a great deal of support which in turn have made the traditional heads of the Khasi *himas*, the heads of *Raids*-conglomeration of villages and *Rangbahshnong*- village headmen almost indispensable within the structure of administration. This in turn has made increasing interaction and dependence on the traditional institutions by the Khasi Hills Autonomous District Council and the Shillong Municipal Board. Indeed there are situations where citizens have experienced that they have for some reason or other to get the sanction of one, two or even all three of the traditional institutions, be it a residential certificate, opening a bank account and such requirements. And all this and more from institutions that are yet to get constitutional recognition.The paper will make an attempt to study in some broad detail the position of the traditional institutions in the region and bring the discussions to more contemporary times. The canvas will be broad but will eventually narrow to the Khasi situation as we see it today. Reference will be made in the historical narrative to the colonial pattern of administration and the changes that came after independence.

The North East region became part of the Indian state as a consequence of British rule and given an Indian identity through the last two centuries. This process began in 1826 with the acquisition of Assam, Cachar in 1832, Jaintia in 1835 and the annexation of the hill periphery through the 19th century closing with the annexations of the Naga and the Lushai Hills.The Assam plains districts were regulated districts, administered just as other parts of India. The hill areas in the region that came under the direct control of the British colonial state were categorised by the Government of India Act 1935 as either Excluded Area or Partially Excluded Areas. The Excluded Areas[1] were under the executive control of the Assam Governor. The Partially Excluded Areas[2] were under the control of the Governor and subject to ministerial administration, but the Governor had an overriding power when it came to exercising his discretion. No act of the Assam or Indian legislatures could apply to these two hill divisions unless the Governor in his discretion so directed. He was empowered to make regulations for the hill districts, which had the force of law. The administration of these hills was his 'special responsibility'. With no representatives in the Assam Assembly (other than the Partially Excluded areas, which sent one legislator each), political activity above their village and local levels before 1947 appeared to have been in a nascent stage other than in some urban settings.

Within the region were the twenty-five Khasi states-*himas*, Manipur and Tripura. The Khasi states were nominally under the administration of their chiefs (*Syiems, Lyngdohs, Sirdars* and *Wahadadars*) in *Dorbar* with only a supervising attention from the Deputy Commissioner of the Khasi and Jaintia Hills District who doubled as their Political Officer. Much of Manipur State comprised the hills around the Imphal valley. The state's Political Agent, stationed at an impressive Residency in the heart of the capital town was vested with certain special responsibilities in respect of the administration of Manipur hill people. Tripura never appeared to have had a tribal policy despite the tribal *Maharaja*.

There was a third area inhabited by tribals in the North East. Along the watershed between Burma and further north in the hills reaching the eastern Himalayas live Naga tribes, Miris, Monpas and a kaleidoscope of other kindred peoples who were 'unadministered' as the official termed such areas outside their control. The British in punitive expeditions and survey operations only occasionally visited their hills.[3]The Naga Tribal Area and the Tirap Frontier Tract were technically and for practical purposes outside British India. There was a statutory boundary between these two frontier tracts and the adjoining districts of the province.[4] While this boundary had been defined, by an oversight admitted by the British administration, no similar notification was issued for the northern boundary of the Assam province as it was assumed that the whole territory up to the Indo-Tibetan frontier was *de jure* an excluded area and so theoretically formed part of the province.[5] However, this interpretation was contrary to the administrative position because while the Government of India treated the area as tribal and unadministered, treaties of 1862 and 1874 with the tribes of these hills refers to them as foreign with a distinction made between "the boundary of the Queen and your country" and the limits of British territory was fixed at the foothills.[6]

The three categories of hill areas referred to above were integrated into the Indian State shortly before and in the months following August 1947. The British India areas were automatically made part of India while the 'native' states with some tact and intimidation were integrated and merged with the Indian Union.[7] Composite Assam then included in its jurisdiction much of the hill areas referred to above. A number of factors which we cannot at this point enter into discussion were responsible for the breakaway of the hills from Assam. The process began with the emergence of the Nagaland hills in 1963. The 'Hill State' political activity of the East India Tribal Union and the All Party Hill Leaders Conference and their agitation against the language bill of the Assam legislature to impose the Assamese language on the non-Assamese speaking hill people were the significant developments of the 1960s-1970s. The political activity ushered in the state of Meghalaya and the Union Territories of Mizoram and Arunachal Pradesh in 1972. Earlier both Tripura and Manipur, which were Union territories, were upgraded to full-fledged states in 1972. Mizoram and Arunachal have in time become full-fledged states.

Non-Regulation

The Non-Regulation pattern of administration was operative in the hill region. First applied to the Garos by Regulation X of 1822 and later elaborated by a series of official administrative measures including the Inner Line Regulation of 1873 and the Scheduled Districts Act 1874, the system was characterised by a simple procedure of administration. The full provisions of law and administration were not made applicable to the tribes inhabiting these hills. In time the Non- Regulation system was adapted for the tribes of North East India as each were brought under British rule; and later extended to the tribals of Chota Nagpur, Jalpaiguri, Darjeeling, the Chittagong Hill Tracts of Bengal, Kumoan in the North West Province and in certain districts of Sindh.[8]

With their subjugation by the British, steps were taken by the colonial administration to give to these hill people a 'paternal' government that allowed them to exercise their own genius in the management of themselves, with just that amount of control from above. Traditional leaders were encouraged to continue their age old and traditional administration under the new dispensation. Their traditional hierarchical structure remained. The colonial administration permitted their chiefs to continue their authority over their respective villages under the watchful eyes of the district authorities. In the British India villages, the Khasi *Sirdar*, the Naga *Goanbura*, the Mizo *Lal* and the Garo *Nokma* assisted the administration in collection of revenue, house tax and other functions for which they were paid a share of the collection. Naga *Goanburas* were each given a distinctive red blanket as a symbol of their authority. This form of indirect administration operated in the hills whereby the chiefs became the local props; if such a term is appropriate, to describe the relationship between the rulers and the traditional heads of villages.

British rule over the hill people of the North East superceded centuries' old political isolation of the hill people and introduced a pattern of administration to suit their policy. An important feature of this administrative pattern was the integration of distinct tribal areas into a district or sub-division named after the predominant tribe. In the process of acquisition and consolidation of British administration over these hills several of their indigenous institutions and customs were conveniently allowed to become defunct. On the other hand certain other institutions and laws were introduced which were alien to tribal traditions. Generally however, and in line with the British policy of non-interference with the traditional forms of governance of the people, customs in their life and culture and governance got sanction without being spelled out in any detail

The politics of the Indian National Congress, the All India Muslim League and other political parties of the Assam legislature had minimal effect in these hills. The British policy of exclusion / segregation of the hill tribes kept them out of touch with the political, cultural and other developments in other parts of India. However it was not long before political consciousness emerged in the hill districts. It started with the Jaintia Dorbar in 1900. From the second decade of the last century political activity in these hills increased in view of the constitutional reforms Britain offered India. The participation of several hill tribes in the war of 1914-18 exposed them for the first time to peoples and places far from their homes. Naga political activity began with the establishment of the Naga Club in 1918. This was followed by the more hectic political maneuverings of the Khasi National Dorbar set up in 1923 and the Federation of Khasi States, formed in 1934. The pace of political activity increased sharply shortly before the independence of India. The Garo National Council brought Garos together in 1946; the Mizos with popular initiative formed the Mizo Union in April that year.

Modern political activity, as distinct from the traditional governance was not advanced enough in the hills by 1947 to meet the aspirations of the hill people at that crucial time in their history. The integration and merger process, which Indian native states underwent, to become part of the Indian Union, left much to be desired by the people of the region. There was also a lack of appreciating the tribal mind by the new Indian leadership- particularly the Assamese politicians. [9] While some hill leaders were becoming involved in their changing status and position, such Rev. J. J. M. Nichols-Roy, a Khasi and minister in the Assam Government; the people at large were left untouched by the political advance in India. The Indian freedom struggle consequently had little if any impact on the hill people. They had little participation in the struggle. They did have their own struggles against expanding and 'creeping' British imperialism. However the nature and intent of their resistance to British rule was in no way connected to the Indian freedom movement. [10]

Their political integration in India complete, the tribal people in the region were provided autonomous district councils and regional councils as per the provisions of the 6th schedule of the Indian Constitution. Operative in the Garo Hills, Khasi and Jaintia Hills, Mikir Hills, Lushai Hills from 1952 these Councils are provided legislative, judicial and executive powers on certain subjects relating to the tribe(s). [11]

Traditional Governance

Traditional institutions of governance in North East India as we have them today, largely originated among pre-literate communities in pre-colonial times. Consequently the tribes were not able to put in writing the powers and functions of the different forms of governing themselves. The numerous tribes and communities who inhabit the region have oral histories. These have become useful in establishing their history since antiquity for some, to more recent times for others, and even to contemporary history for some migratory groups whose settlement in the hill and plains of this part of India has occurred in living memory. While it may be possible to reconstruct the pre-history of some of the tribes from study of their material remains, it becomes difficult to assign them 'ancient' and medieval' pasts because there is an absence of written material from which to reconstruct their past. Even giving them a 'modern' history becomes difficult because to move into the modern without an intelligent account of what occurred before this period would be contributing to a fallacy in history.[12]

Mention has been made of these problems to put in perspective our understanding of traditional institutions in the North East. For institutions to become tradition a past that is somewhat long in time is required. This past varies for the tribes in the region. Khasi recall their transient settlement in the Brahmaputra valley that was well before the entry of the Tai Ahoms into the same valley early in the 13th century. Meiteis have assigned a fixed time of their settlement in the Imphal valley to the fourth decade of the first century. It is not possible to assign a date to the Naga diaspora into the region and over what are today fours adjacent states, or to more recent movements of tribes such as the Mizo- Kukis whose settlement in their hills is of more recent origin. Historians may hazard tentative dates for these movements and settlements by judicious use of oral traditions cross-checked with other forms of information of a given phase of human history.[13]

It is not certain when and why the institutions that the tribes came to accept as their own originated and under what circumstances. Much of what has been written on their origins is based on conjecture or on material that in all probability cannot be verified as being near to the truth. However, what is to be appreciated is that from their uncertain origins to the present, the institutions have stood the test of time. Inasmuch as the traditional institutions appear outdated and obsolete in the present, they continue to function, not always in the desired form but nonetheless with societal support for both their usefulness and their social control.

The earliest references to the institutions were written by observers who saw in the numerous patterns of tribal government, forms quite different from their own and institutionalized systems of governance.[14] Later, writers from among the tribes both elaborated and clarified the views on their institutions.[15] There thing lay for some time other than the occasional inquisitive academician still in search for answers to his or her questions. Today the study of traditional institutions has assumed significance. The Khasi heads of traditional institutions are seeking constitutional recognition for their institutions in a bid to fit the institutions into the modern system of governance. Several papers have then come up with arguments questioning their functioning and their relevance of which more will be said later in this presentation.

Some Institutions

The tribes developed the institutions of governance out of their own genius and perhaps after many trials and tests till they arrived at a form or forms of governance they considered good for themselves. The tribe that has an experience of three stages of governance, the village, cluster of villages and the state- *hima,* were the Khasi-Jaintias. The basic unit of political organization in the Khasi-Jaintia society was the village, which composed of one or two decent groups. Village administration was conducted by an assembly of all resident adult males under an informal headman elected by them from among their number. When new villages were formed the new community did not detach themselves from the original village but remained an integral part of the growing state or *hima.* Administrative and political necessity led to the institution of tribal leaders such as the *Basan* and *Lyngdoh.* The *Basan* was entrusted the conduct of the clan while the *Lyngdoh* was entrusted both administrative and sacerdotal functions. Under them there were the *Pators, Sangots* and *Matebors* who assisted in the administration of the *Shnong* and the *Raid.* From these rudimentary beginnings it is believed emerged the institution of *Syiemship* which probably arose out of voluntary association of villages when new developments such the opening of markets, marriage laws, organization of land tenure and judicial administration brought in the need of a central and common ruler. The *Basans* and *Lyngdohs* who surrendered their powers as rulers did not forfeit all their powers as they and the founding clans of *himas,* the *Bakhraw,* retained some of their administrative and religious functions, even retaining in some *himas* the privilege to elect their *Syiems.*[16]

There was adaptation in the origins of *Syiemship* as some early *Syiems* were drawn from people of non- Khasi origin. *hima* Shella has four *Wahadadars,* a term borrowed for their chiefs from nearby Bengal. When the British extended their political control over these hills they instituted another functionary; *Sirdars* over certain other *himas* and the functionary in charge of the British villages. Such adaptation as evident in tribal governance in the Khasi Hills was also evident in the Naga functionary, the *Gaonbura,* the Arunachal *Dobashi* and the Garo *Laskar.*[17] This should suggest that while some of these institutions evolved over a long period of time, others are of more recent origin. Indeed the tribes had to adapt and change the nomenclature and functions of some of their functionaries by circumstances over which they might not have had control.

The higher rung of the hierarchical structure (it reminds us that the tribal societies were not egalitarian as it was believed to have been by many writers) of the tribal communities was dominated by their chiefs and assisted by tribal councils. We have made reference to the Khasi- Jaintia structure. The Garos evolved the institution of the *Nokma-* there are four kinds of *Nokmas,* the *Gamni Nokma*; the *Gana Nokma*; the *Kamal Nokma* and the *A'king Nokma* -only the last of the four was entrusted with governing the village.[18] The Garo institutions just as the *Syiem* of the Khasi *hima* of Khyrim are somewhat unique. Whereas the other institutions are patriarchal in nature, the *Nokma* is subject to Garo laws of inheritance through the female line, the *Mahari*; while the *Syiem* of the Khasi *hima* of Khyrim is the son of the *Syiemsad* of that state.

It is not possible here to go into all the institutions prevailing in tribal North East India. It may suffice to take some case studies to explain the nature and function of the living patterns of tribal people governing themselves. The Naga institutions of chief as it exists today have come down with Naga life over a long span of time and under conditions which would have only allowed for strong men as their chiefs. A recent report has summed up the Naga position thus: "The traditional system of Naga polity varies from autocracy (Konyaks), gerontocracy, (Aos, Tangkhuls) and democracy, (Angamis, Chakhesangs, Rengmas, Maos). Among the Semas, the position of the chief is a little less arbitrary than the Konyaks, but is nevertheless highly autocratic."[19] This simplistic explanation of the prevalent situation in the Naga hills inclusive of the Naga inhabited areas within Manipur state overlooks the fact that succession to the position of village chief is usually hereditary except among some tribes. The moot point is in their functioning- while many of the chiefs take support from the adult male population of the village in discussion and consultation through the village councils - the authoritarian Konyak *Angh* and the Sema *Akeko* are not bound to follow their functioning as their kindred chiefs among other Naga tribes.[20]

Whatever may be their structure at the top, the tribal societies put safeguards in their governance through the village councils that performs the work of administering the village, cluster of villages or the larger conglomeration. With their different names they nonetheless have similar functions, administration of justice according to customary laws and practices; keeping watch and ward, arranging the cycle of the *jhum* fields where this practice is prevalent; overseeing the use and distribution of water and other resources; to name some of the functions. Later Khasi *Syiems* would be entrusted by the British administration magisterial duties within their own *himas*. This judicial function continues to be given to the *Syiems* by the Indian administration though with reduced powers for the disposal of civil cases.

What we refer to as traditional institutions today must have been a matter of everyday life for the societies under review. They lived with their own structures under conditions of privilege for certain sections of the society or of oppression and exploitation of others. They fulfilled a role, which at that time had nothing better to offer. There was no alternative. The institutions built up were particularly not fair to their women. They could not aspire for positions of leadership; they were denied entry into the councils and were not heard or consulted. They could not participate as equals to their men-folk in these tribal societies. The literature that is now available on the position of women in the traditional tribal societies suggests that tribal women were not as exploited as their counterparts in mainland India - but it does not speak well of the tribesmen to have given secondary position to their women.[21]

Debate on Traditional Institutions

In recent years academicians have focused much attention on the traditional institutions of governance in North East India. Numerous seminars, conferences and workshops have discussed in detail the powers and functions of the institutions relating to particular tribes and communities; their management structures, their relevance, and other related issues.[22] The literature generated on the subject is increasing and varied.

In more recent years the focus has shifted from the descriptive to an analytical and questioning of the institutions. Five papers of very many other presentations will be discussed. The Crisis States Programme of the London School of Economics and Political Science intends to provide new understanding to the causes of crisis and breakdown in the developing world and the processes of avoiding or overcoming them. As part of the programme a working paper has been prepared on North East

India.[23] Apurba Baruah must find it convenient to work on the tribal traditions in Meghalaya as the material is easily at hand. Ably presented, he has applied a critique of the institutions at the local level without an appreciation of their functions and their relevance to the Khasi society as the community sees these institutions. He begins the paper with references to traditional institutions in Africa and the North East; the question of their tribal status; and then reviews what scholars have written on the subject of traditional institution in Meghalaya, with no mention of the Garos. Towards the conclusion Baruah says that the most influential of traditional institutions are to be found in the *Dorbar shnong* level in Meghalaya! (read Khasi- Jaintia Hills) Allowing the insensitivity of the author to his subject of study, the paper, very quickly concludes without any suggestions that an understanding of the *Dorbars* in rural and urban setting may help resolve the conflict of values that otherwise may create a major crisis of governance.[24]

A second paper is in circulation on the same subject.[25] Bengt Karlsson's understanding of the subject he writes makes interesting reading. Though the paper is polemical in content, the author has much more sensitivity of the subject of traditional institutions. Karlsson seeks to explain the issues on ethnic homelands and indigenous governance in the region, again with reference to the Khasi- Jaintias, not so much from their past but from more recent factors. He links the demand for recognition and empowerment of the traditional institutions in part to the poor performance of modern institutions of governance. He makes mention that (a section of) the Naga favour a return to indigenous customary institutions in place of party politics. Cautious in his conclusions, the scholar sees the indigenous governance as "a worrying sign that spell further cleavages and ethnic conflicts". On the other hand the author sees the possibility of the traditional institutions with a more locally grounded form of politics. He sees a need that the traditional councils be reformed and made more inclusive in order to get democratic legitimacy. Should this be possible, Karlsson is of the opinion that there can be a widening of civic space in Meghalaya.[26]

Policy making bodies have become involved in these studies. The National Commission to Review the Working of the Constitution has published a Consultation Paper on the subject.[27] Intended for the sole purpose of generating a public debate and eliciting public response, the Consultation Paper has provided an overview of the institutions operative in the hills and plains of the region. The outcome of this exercise is not known, thought the historical background, questionnaire and appendices of the Consultation Paper have become useful material to build up a critique on the traditional institutions. The question may be asked what has been the outcome of the Consultation, the papers it commissioned and the cost of its operation?

Among several writers of the Khasi community who have written on the subject the articles of Toki Blah and Charles Reuben Lyngdoh are brought into this discussion. Blah's essay appeared in a popular daily.[28]Traditional institutions in rural Khasi Hills have not undergone much change. If there is a perceptible change in the functioning of these institutions it is in Shillong where the headmen or *Rangbahshnong* (more than the *Syiem* of Mylliem, within whose *hima* they function) who have had change to their status and functions. Blah begins his essay with the respect the community has for the institutions. He asks the question how to make use of the institutions and how adaptable are they to our modern needs? He sees a possibility to remodel the traditional institutions and strengthening them with new technology, as the luxury to build new viable platforms is not with the community. He argues that it would have been in the interest of the community were the *dorbars* entrusted a greater measure of governance of the Shillong rather than the Shillong Municipality electing Commissioners of Wards. "Staid unwritten political acumen was unwittingly exchanged for innovative untried political stratagem." He feels that the *dorbars* could have been facilitated, trained and capacitated to take up responsibilities within an urban setup. The author has traced the failure of the elections to the Shillong Municipal Board and explained how it was that the traditional headmen have stepped in to assist the Corporation in its functions. As he sees it, today each *dorbar* functions within its own limited jurisdictions. As regulatory bodies they have been found to ably assist the civil administration in several functions. There is however no coordination between these *dorbars* other that an overarching *Synjuk Ki Rangbah Shnong*. He concludes the two-part essay with an appeal to recognize the potentials of these self-governing bodies and orienting them towards closer cooperation with the administration.

While the essay noted above sees a role of the traditional institutions in the administration of urban Shillong, the presentation of Charles R. Lyngdoh[29] poses the question whether the Khasi *himas* are ready and have the capability to undertake developmental work. Earlier the *himas* survived with little or no funds. Today funds from the Central government and from grants under the discretion of Members of Parliament and Members of the Meghalaya Legislative Assembly are being made available for the *himas*. Further, encouraged by the setting up of the National Commission to Review the Working of the Constitution in January 2000, the Himas have suggested that they may come under a new provision of the Constitution a13[th] Schedule, to provide for a Federal Council with legislative, executive and judicial powers and with direct funding from the Centre. The experience of the traditional heads of the *himas* with the Khasi Hills Autonomous District Council since its inception has not been cordial. They continue to grouse against the 1959 act of the Council that lowered their position to that of village headmen![30] With the changing role of the *himas* and their institutions, the scholar advocates a thorough review to make them more vibrant and in tune with the requirement of civil society. He suggests wider participation of youth, women and professionals in the developmental activities in the Khasi states.

Conclusion

The institutions at all three levels, briefly reviewed above are now faced with a dilemma - to change in part or substantially and to further adjust into the situation within which they have survived. Faced with these problems these institutions are making both bold attempts and feeble bids to adjust and change. They require changing or becoming set, rigid and moribund. In part the hardening of the traditional leaders at the *hima* level, to the changes they face comes as a reaction to the unsympathetic if not unconcern the Indian State gave them over the past five decades. The Indian State lost sight of the role traditional institutions could have played in the administration of the tribal people in a more effective manner. It did not take the lessons from the British who had used the chiefs and their councils as useful instruments in what has been called 'indirect administration.' Rather than utilise the experience of their leadership, a new institution in the District Councils was introduced under the Sixth Schedule of the Indian Constitution, composed of elected representatives of the tribal population to oversee as part of their functions, the working of the age-old traditional forms of government. With the introduction of the District Councils the situation has actually deteriorated to mistrust of one institution for the other; a misuse of power by the newer form of administration which has in most cases failed miserably. It is no surprise therefore that we are witnessing in our own time a resurgence of the rights and powers and privileges of the traditional institutions and their leaders.

Caution should be the watchword of the traditional institutions as they seek constitutional recognition. Times have changed. For instance not all Khasis in Shillong, which falls within the *hima* Mylliem, accept the *Syiem* of Mylliem as their *Syiem*. Khasis residing in the capital of Meghalaya continue to have connection and affiliation with the village and *hima* their families originally came from. There are strong reservations to the continuation of the traditional institutions when there are newer and by now time tested forms of governance that guarantee equality before the law, give women their rights and stand for accountability.

As more funds become available for development and the State administration entrusts more power to the traditional institutions at the local level, the institutions have been swayed away from their moorings into the modern political system and administration. The 'non- traditional' leaders within the community for their own political gain may use the dependence on the traditional institutions to have a hand in development of roads, education, water supply and infrastructure within their respective jurisdictions. The modern state has the funds, the traditional institutions do not and here lies the bait. The relationship now cannot be one of 'indirect rule'. In their new role it may be possible to provide the institutions a larger share in the administrative processes in a way that will ensure their participation. If they are to continue, they need have the foresight to adjust by such measures as providing women a place in the institutions' participation and decision-making, including accountability and transparency in their management and accounts.

Caution should also be the watchword of the authorities that will have to respond to the pressures for greater powers and constitutional recognition being sought by traditional institutions. This demand had arisen in the background of the anxiety the traditional heads of Khasi *himas* face with the Khasi Hills Autonomous District Council. If the District Councils have failed, there has been no official admission until very recently that the Sixth Schedule which provided for the Councils was "institutionally defective and deficient" to meet the aspirations of the hill people.[31] When the Nagas found no use of this institution and continue to hold on to their own traditions was it necessary to impose by legislation such councils on other peoples at such cost and little good? On the other hand the question may also be asked and answers should be attempted to explain why, if the Mizos did away with their institution of *Lal* so soon after the independence of India, and have not been any worse off, why was it not possible for other tribal people to have changed if not done away with traditional forms of governance?

Two crucial issues have not been discussed in any forum. The Khasi traditional forms of governance were devised when the tribes lived in relative isolation and for themselves. Today the tribes live in close proximity and in the larger urban centres, together and close to peoples of other communities. When peoples live so close to each other how then can there be one set of governance for one section in which others are completely exempted? For instance, one Saturday in late 2003, the Laitumkhrah Dorbar called for a meeting of all Khasi residents of that *shnong*. This meeting put restrictions on the movement of people in and through Laitumkhrah, a commercial centre of the city of Shillong and caused tremendous inconvenience to many others. Secondly, if the tribes set up the institutions, adapted and were given institutions, which have become theirs, then it should only be to the respective tribes to take the decision whether they can live with or without them.

There are many and overlapping levels of governance for the tribal people. The functions of their traditional leaders even today in the modern state coexist in a curious mixture of the traditional and the modern. The traditional governance of the village headmen and elders continue to function in tribal areas of the region. As archaic as they appear and are, they are nonetheless respected and functional. They know the limits of their functions. Rumblings of discontent are beginning to be heard, particularly from the educated and urban sections of the community, of the usefulness of tradition forms of governance when the Indian state has provided mechanisms of modern and democratic governance.[32] Presently a debate is underway in the Khasi and Jaintia Hills following the decision of Justice S. R. Sen that village headmen- *Rangbahshnong* are not authorized to issue 'no objection certificates. The debates come out of the increase of the powers and functions of the *Rangbahshnong*, over the years, whose supporters draw on the continuation of tradition and usage to support its continuance. The younger intellectual and leaders of civil society question the authority given to the village headmen, many of who have clung to their positions for years and without any election by all adults of the village concerned. There are also issues of gender bias in the process of election if any, and the participation of women in decision making in the villages.[33] It would be fair to say in conclusion that this state of affairs has arisen because of the acceptance of the necessity for various reasons, to preserve the traditional institutions within an evolving Indian democracy. In all this the understanding of "tradition" within the "modern" will continue to be a subject of debate and policy.

Notes and References

*David R. Syiemlieh, 'Traditional Institutions of Governance in the Hills of North East India; The Khasi Experience', *Man and Society, A Journal of North East Studies*, VOL. III, Indian Council of Social Science Research North Eastern Regional Centre, Spring, 2006, pp. 117-137.

1.   The Excluded areas comprised the Naga Hills District, the Lushai Hills district and the North Cachar Hills District.

2.   The Partially Excluded areas were the Khasi and Jaintia Hills District, the Garo Hills District and the Mikir Hills.

3.   Refer to J. P. Mills, *The Pangsha Letters: An Expedition to Rescue Slaves in the Naga Hills*, Pitt Rivers Museum,1995. This is a collection of letters of Mills during the 1936 expedition.

4. India Office Library and Records, L/P&S12/3115A. No. 22, File 6, Memorandum on the Tribal and Excluded Areas of the North Eastern Frontier, para 4.

5. *Ibid*. Cambridge South Asia Archives, Pawsey Papers, Box 1, No. 5, Andrew Clow to Charles Pawsey, 24 April 1947.

6. *Ibid.*

7. David R. Syiemlieh, 'The Political Integration of the Khasi States into the Indian Union,' B. Pakem (ed.), *Regionalism in India*, Har Anand Publications, New Delhi, 1993, pp.147-156; Sajal Nag, *India and North- East India: Mind, Politics and the Process of Integration 1946-1950*, Regency Publications, New Delhi, 1998. Refer also to the *Proceedings of the North East India History Association* First/Third/ Fifth and Sixth Sessions for papers on the integration of Manipur.

8. For details of this pattern of administration refer to J.B. Bhattacharjee, *The Garos and the English*, N. Delhi, 1978 and D. R. Syiemlieh, *British Administration in Meghalaya: Policy and Pattern*, N. Delhi, 1989.

9. For a comprehensive understanding of the process of the integration of the hill areas of North East India into the Indian Union refer to David R. Syiemlieh, *Response of the North East Hill Tribes of India Towards Partition, Independence and Integration: 1946-1950*, Pratibha Devi Memorial Lectures 2003, Gauhati University, pp10-27. Also read the chapter 'Genesis of Hill Politics,' in S. K. Chaube, *Hill Politics in Northeast India*, Orient Longmans, reprinted 1999 and the chapter "Final Bid for Power " in Amalendu Guha, *Planter Raj to Swaraj: Freedom Struggle and Electoral Politics in Assam 1826-1947*, People's Publishing House, reprinted 1988.

10. D. K. Fieldhouse in *Economics and Empire 1830-1914*, Weidenfeld and Nicholas, London 1976, pp.80-81,173-175, has used the term "colonial-sub-imperialism" to describe the expansion of European imperial powers in Asia. The term aptly describes the nature of expansion of the British into North East India.

11. S. C. Chaube, op. cit., pp. 100-116.

12. For a discussion on the periodisation of the histories of the tribes of North East India refer to Amalendu Guha, "Introduction', in J. P. Singh and Gautam Sengupta (ed.), *Archaeology of North East India*, N. Delhi, 1982, p.2 and David R. Syiemlieh, 'Technology and Socio- Economic Linkages of the Khasi- Jaintias in Pre-Colonial Times', Mignonette Momin and Cecile A. Mawlong (eds.), *Society and Economy in North East India*, Vol I, New Delhi, 2004, pp. 22-23.

13. There is a plethora of literature on the tribes of the region. Early in the 20 [th] century, the British Government commissioned a series of monographs on several of the tribes. These were useful references on the communities and have in time required both revision and updating. A useful introduction to the tribes may be read in Frank M. Lebar *et al., Ethnic Groups of Mainland Southeast Asia,* Human Relations Area Files Press, New Haven, 1964, has entries on the Nagas, Chins, Garos, and Khasis.

14. A detailed account of the Khasis was written by Alexander Lish and published in the *Calcutta Christian Observer*, March 1838, pp.128-143. An early account on the Garos was prepared by John Eliot. 'Observations on the inhabitant of the Garrow Hills made during a Public Deputation in the Year 1788-89,' *Asiatic Researches*, Vol. III, 1792,pp21-45. On the Nagas and several other hill tribes there are numerous references to them in the *Journal of the Asiatic Society of Bengal* . Brahmaputra Studies Database has placed several articles published in the Asiatic Society of Bengal and other journals on its portal.

15. Useful studies by tribal on their own people are those by Milton S Sangma, *History and Culture of the Garos,* Books Today, New Delhi, 1981; M. Alemchiba, *A Brief Historical Account of Nagaland,,* Naga Institute of Culture, Kohima, 1970; Lal Biak Thanga, *The Mizos: A Study in Racial Personality*, United Publishers, Gauhati,1978; David R. Syiemlieh, *British Administration in Meghalaya: Policy and Pattern*, Heritage Publishers, New Delhi, 1989; Hamlet Bareh, *The History and Culture of the Khasi People*, reprinted Spectrum Publications, Guwahati, 1985.

16. P. R. T. Gurdon, *The Khasis*, reprinted, Low Price Publications, Delhi, pp. 62-75; Hamlet Bareh, op. cit., pp.234-284.

17. *Goanbura, Dobashi* and *Laska*r were terms not earlier used by the concerned tribes. These terms were applied following British rule/ political control over the tribes.

18. Milton S. Sangma, op.cit., pp. 60-66.

19. National Commission To Review The Working Of The Constitution, *Consultation Paper on Empowering and Strengthening of Panchayati Raj Institutions/Autonomous District Councils/ Traditional Tribal Governing Institutions in North East India*, December 2001, p.71.

20. A. S. W. Shimray's, *History of Tangkhul Nagas*, Akansha Publishing House, New Delhi, studies the institution of the Awunga among the Tangkhuls. Chapter 7 summarises the forms of governance among the other Naga tribes.

21. On the emerging research of the position of women in the region read F. S. Downs, *The Christian Impact on the Status of Women in North East India*, NEHU Publication, Shillong, 1996,;Ira Das, 'Status of Women: North Eastern Region of India versus India,' *International Journal of Scientific and Research Publications,* Vol. III, Issue 1, January 2013, pp1-8; Ruth Lalsiemsang Buongpui, 'Gender Relations and the Web of Traditions in Northeast India,'*The NEHU Journal*, Vol XI, No. 2, July 2013, pp. 73-81; Anju Vyas and Madhu Shri (comp.) 'Women of North East India A Bibliography,' CWDS Library Resources Series; XII, Centre For Women's Development Studies, New Delhi.

22. Among the earliest of studies on the traditional institution in the region are S. K. Chattopadhyay (ed.) *Tribal Institutions of Meghalaya*, Spectrum Publishers, Guwahati, 1985. A broader study then followed in Jayanta Sarkar and B. Dutta Ray (eds.), *Social and Political Institutions of the Hill People of North East India*, Anthropological Survey of India, New Delhi, 1990.The Rajiv Gandhi Foundation, New Delhi sponsored a seminar on 'Traditional Self-Governing Institutions,' in Shillong, 29-31 August 1994. L. S. Gassah's study, *Traditional Institutions of Meghalaya; A Study of Doloi and His Administration,* Regency Publications, New Delhi, 1998 has given attention on one of the institutions. Bhupinder Singh has edited *Antiquity to Modernity in Tribal India: Tribal Self- Management in North- East India*, Inter-India Publications, New Delhi, 1998. This is the second volume of a four part series on tribal studies in India. A number of articles on the subject have been included in George Mathew (ed.), *Status of Panchayati Raj in the States and Union Territories of India 2000*, Institute of Social Sciences, New Delhi, 2000. The North East Network, an NGO working in the region, turned attention to the traditional institutions of the region during the Ford Foundation celebrations in India, at Guwahati, in March 2001. A brief report of the meet is published. Atul Goswami has edited *Traditional Self-Governing Institutions Among The Hill Tribes of North - East India*, Omeo Kumar Das Institute of Social Change and Development, New Delhi, 2002. More recently the North East Foundation, Guwahati held a seminar on 'Constitutional Safeguards for the Indigenous People of North East India,' Shillong on 24 November 2001. This was followed by a number of presentations

on the subject of traditional institutions presented at the Department of North East Region, Government of India sponsored seminar " Development Plans for North Eastern Region: Exploring Options and Possibilities", at the North Eastern Hill University, 11-12 December 2003.

23. Apurba K. Baruah, 'Tribal Traditions and Crises of Governance in North East India, with special reference to Meghalaya,' March 2003, Crisis States Programme, Working Paper, No22. Baruah has followed this paper with a more detailed study of Laitumkhrah Dorbar, Shillong. See 'Ethnic Conflicts and Traditional Self Governing institutions a Study of Laitumkhrah Dorbar,' Crises States Working Paper, No. 39.

24. *Ibid*.pp. 8-10.

25. Bengt G. Karlsson, 'Ethnic Homelands and "Indigenous Governance": Reviving 'Traditional Political Institutions in Northeast India,'.Draft paper for private circulation. Another of Karlsson's papers on the subject and recently published is 'Sovereignty through Indigenous Governance: Reviving 'Traditional Political Institutions' in North East India,' *The NEHU Journal*, Vol. III, No. 2 July 2005, pp.1-15.

26. *Ibid*. The paper has no pagination.

27. National Commission to review the Working of the Constitution, Consultation paper on *Empowering and Strengthening of Panchayati Raj Institutions/Autonomous District Councils/ Traditional Tribal Governing Institutions in North East India* , December 2001,http://ncrwc.nic.in

28. 'Traditional Institutions and Urban Governance,' *The Shillong Times*, 3 and 4 May 2004.

29. 'The Politics of Development: Contesting the Claims of the Himas', Paper read at the seminar 'Development Plans for the NE Region_ Exploring Options and Possibilities', NEHU, 11-12 December 2003.

30. Memorandum Seeking Recognition and Protection of the Traditional Institutions of the Khasi Race, submitted to the President of India, Steering Committee, Dorbar Hima Mylliem, 28 February 2001; The National Commission to Review the Working of the Constitution, Consultation Paper on *Empowering and Strengthening* etc, op. cit. appendices I and II.

31. *The Asian Age*, 2 June 2002.

32. Patricia Mukhim, 'Speaking out of Turn', *The Shillong Times,* 23 July 2004.

33. The Judge observed on 10 December 2014 that headmen "have no power to issue any kind of certificate unless empowered by rule or laws". The judgment and order was passed when the judge was adjudicating on a writ petition filed by eight residents of different villages of East Jaintia Hills who wanted their fundamental rights to be protected. Andrew W. Lyngdoh, 'HC redefines headmen power', *The Telegraph*,17 December 2014; Patricia 'Mukhim, 'Of High Court Ruling, KHADC and Rangbah Shnong and Tradition', *The Shillong Times*, 13 February 2015. The State government has informed that it would formulate its own laws to define the role of headman, *The Shillong Times*, 21 February 2015.

17

The North East: Home of many Tribes and Communities

There are few places in this world where such a variety of peoples live in so close proximity to each other as in North East India. The demographic composition of its inhabitants range from those of the Brahmaputra and Barak valleys who are predominantly Assamese — descendants of the pre-Ahom settlers including the Bodos and related tribes, Ahoms who migrated and settled in the valley and Muslims following Mughal interest in Assam and migrants from beyond. The Barak valley being an extension of the Bengal plains is peopled by Bengalis of both major beliefs – Hinduism and Islam. Tripura was largely populated by tribals. More recent demographic changes has resulted in its indigenous and tribal people being outnumbered by migration and settlement of people largely from east Bengal. The generic term 'Arunachali' which is of recent usage, groups together people of the new state. The Nyishi, Galo, Shedrukpen, Monpa, Mishing, some Naga tribes, Singpho and other tribal groups of this once inaccessible and remote stretch of land are racially Mongoloid and speak languages/dialects of the Tibeto-Burman group — this is the single most important link of the majority of the tribes which spreads from Arunachal Pradesh round the Naga Hills and to the south into Mizoram and part of Tripura.

The sixteen major Naga tribes living in the hills after their name have a long tradition of migration and diaspora into neighbouring Arunachal Pradesh, Assam and Manipur apart from a large group living across the frontier in Myanmar. The people identifying themselves as Mizos have taken the name of the dominant group in the state of Mizoram. These include, apart from the Mizos themselves, the Lushai, Hmar,

Paite, Pawi and other Kuki people. Manipur with its Meitei population in the Imphal valley and a small pocket in Jiribam close to Silchar, is surrounded by Naga and Kuki tribes, who having lived for ages in their own settlements, are today in a situation of anxiety with neighbouring tribes and communities.

The generic terms referred to earlier and the use of clan cluster concepts such as Zomis and acronyms such as those of the Zeliangrongs and Chakesangs are of recent developments. At the time of their migration into the hills of their present residence each tribe had its own identifying marks in a common past, common heritage, beliefs, language and area of occupation. They even had distinct physique that anthropologists find easy to differentiate. It was the British who gave the tribes their group name such as Naga, Lushai, and Khasi, to name a few. The tribes, however, prefer to keep to their own group of Ao, Sema, Mizo, Khasi or Pnar. It is not certain when these tribes migrated into their present locales. That they were or were not autochthons is still being researched. They came in several waves over a period of centuries. Most of the tribes, the Ahom included, trace a migration from further East. The Naga culture appears to have been a remnant of an ancient civilization which in pre-historic ages extended over a large part of South East Asia.

Meghalaya

Within this very complex but colourful and vibrant group of people who call the hills and plains of North East India their home are the matrilineal Khasi-Jaintias and Garos of Meghalaya, who will be the focus of this essay.

The history of the Khasi-Jaintias and Garos before the arrival of the Ahoms into their hills is uncertain. Khasi folklore and oral traditions tell about their supernatural origins, which a section of the tribe have interpreted to explain that they are autochthons of the land now called Hynniewtrep. This belief apart, it is generally held that they were one of the first tribal groups to have migrated into their present hills, though when that migration took place and why there was a migration has not been explained. However what is certain is that this migration preceded that of the Ahoms, as *Ahom Buranjis* make no mention of the Khasis in the course of their own migration into the upper Brahmaputra valley, whereas there are references to the Nagas and other peoples the Ahoms interacted with in the course of their settlement in the valley.[1] The Garos who identify themselves as A'chik or Mande, are a branch of the Tibeto-Burmans. Related to the larger Bodo stock including the Hajongs, Reangs, Mech and other kindred tribes, their traditions speak of a migration from Tibet and then eastward into the Assam plains and its surrounding hills. The Garos largely settled in the hills that have taken their name. Their diaspora however, covers an area as far east as north Bengal, the plains of lower and middle Assam, towards the Mikir hills, Mymensingh and Sylhet and in more recent times migrating into Tripura.[2]

When and where the Mongoloid Khasis changed their language to its Austric base is another of those unsolved feature of their past. Their tradition says that they lived for some time in the Brahmaputra valley from where they settled in the hills that have come to take the name of the tribe. They are said to have first settled in the Jaintia hills moving gradually toward the Khasi hills in the practice of swidden cultivation and search for iron ore. That the tribe was one and whose roots are common is explained in their genealogy and clan structure. The Diengdoh clan of Sohra trace their origin to *ka* Iaw Iaw, whose mother came from beyond the Kopili. One branch of the clan resided in Jowai where it became known as the Laloo clan. Another branch went to Nongkhlaw and became the Diengdoh Kylla clan. The fourth branch went to Mawiong and became the Pariong clan.[3] The story of this clan, to take only one example is remarkable for in part it points to an eastern point from where the tribe came as also it explains the formation of the Khasi clan structure. In all likelihood the tribe was spread over a wide geographical area; apart from the Meghalaya plateau they are reported to have resided in Kamrup where Ahom records mention the names of some of the Khasi *himas*, and south into Sylhet where much of the lowlands toward the river Surma was in the possession of the hill chiefs.[4] After what could have been a long time the Garos took to their hills. Mughal and early British records relate the control of the outer fringe of the Garo hills by five Zamindars of East Bengal, a condition that the British corrected in the 1860s.[5]

With histories that go into antiquity, and hemmed in by Goalpara in the west and the Kargi Anglong in the east, with Goalpara and Kampur to its north and Sylhet and Mymensing in the south, the Khasi–Jaintias and Garos were largely confined to their hills where over many centuries of habitation the Khasi- Jaintia and Garo ethos took shape. When and how these two communities first came up to their hills and became matrilineal societies remains unsettled.

State/Village Structure

Similar to tribal societies in other parts of the region, the Khasi-Jaintias and Garos inhabiting the Meghalaya plateau and its surrounding region are closely connected by ties of affiliation to their villages and larger territory, language, blood relationship, belonging to a clan, endogamy, culture, religion and political organisation. These are several of the discerning factors for belonging to and indentifying with a tribal society

It is uncertain when and why the Khasis and Garos left the Assam plains after their migration into the valley. The Garos had evolved their village and social formation perhaps much before their occupation of the hills. The Garo system was a typically tribal formation in which clan based units called the Machong was the unit of social, political and economic life of the tribe. In the process of taking possession of the hills the Khasis evolved their own social and state structure. They first moved in the southern direction into what was to become the Sutnga *hima* ; there from they moved both south into the plains of the Surma valley and west into the central highlands of old Shillong and Sohra *himas* and further towards the westernmost part of their settlement in the *himas* of Nongkhlaw, Rambrai, Nongspung, Jirang and Nongstoin. Though there is no sequence known of this migration and settlement, that the settlement was from east to west can be substantiated by the advance of their technology in erecting of monoliths and the dressing of some of the larger stone remains and the oral tradition that the move westward was in part triggered by the search for iron.[6]

The state formation process was relatively far advanced in the Jaintia *hima*. An element nurturing the more developed state formation process in Jaintia, the easternmost of the "Khasi" states was the Brahmanical myth of the origin of that *hima* which J. B. Bhattacharjee has studied in detail.[7] This apart it was also the only hill state to have had control of a large part of the Sylhet plains beyond the foothills, extending even up to the river Surma. Another factor for the more advanced nature of the Jaintia state was the size of this *hima*. It abutted on the Karbi hills further east, into the foothills in Nowgong and Kamrup, and running across the entire hill region of the Meghalaya plateau.

State formation in other Khasi *himas* was uneven. Tradition tells of the existence of thirty Khasi *himas* and twelve *Dollois* in pre-colonial times. The state formation process of these *himas* was somewhat different to Jaintia, in that many were very small to small *himas*; not all were directly in contact with either of the peoples of the Brahmaputra and Surma valleys and not all had Brahmanical influences as Jaintia and some of the *himas* in the *War-* the southern region were to experience.

Linkages

The two tribes lived in relative isolation for generations, with only the occasional interaction between themselves and with the people of the surrounding plains. Their connection with the communities from where they are said to have migrated was broken if not tenuous. Located in the fastness of their hills, they were drawn to have a much closer connection with the larger Indian state and its cultures. British imperialism in its many dimensions and its 'add on', Christian missions, provided this stimulus.

However, much before British expansion and the completion of empire there were religious, social and economic linkages between the North East and mainland India. These linkages go back into antiquity. Buddhism was known to have had some impact in the region. The Hindu tradition is long and profound. Kamakhya and Jaintiapur[8] have been places of religious significance from times past. Assam underwent social and religious movements the effects of which are evident today. The impact of Islam was more recent yet with significant following. All these religious features have had a profound sense of belonging to the larger Indian ethos. The Brahmaputra and Barak valleys were further opened to migration of ideas, institutions and people from the mainland from times past. Economic linkages and markets further added to this sense of a larger and imagined nation. The waterways and settled population made it possible for a number of linkages to be established. All these have remained and become part of the regional yet distinct variation[.9]

The situation was somewhat different with the several hill tribes in the region. Their interaction was largely confined with the people around and over their hills and the plains below with whom there were more of barter-economic relations. The larger states in the valleys did not venture to have more than cordial relations with only occasional raid into the hills. It was in this situation that these hills were incorporated into British India.

Of the several traits of the Khasi-Jaintias and the Garos that are clearly evident three stand out, their matriliny, their traditional institutions of governance and their stone culture.

Matriliny

Khasi-Jaintia and Garo women of Meghalaya have a much higher position than do their counterparts in other, including tribal societies. This in large part is due to their social structure. The Khasi-Jaintia and Garo are governed by matrilineal principles of descent and inheritance.[10] In line with this more ancient tradition of lineage, children take their mother's name, they belong to the mother's clan and inheritance of ancestral property descends from mother to daughter. They generally follow the uxorilocal pattern of residence. Regional variations prevail as among the *War* in the southern foothills. Changes are taking place in inheritance. One has to be a Khasi-Jaintia or Garo to fully understand and live according to the clan and social norms- not that there are no detractors from within their societies who want change for a different social structure.

Governance

After their own genius the tribes developed institutions of governance which they considered good for themselves. The Khasi-Jaintias evolved the institution of *Syiemship* for their *himas* with cluster of village- *Raids* and villages -*Shnong* under the *hima*. Other institutions within the structure included the *Basan, Lyngdoh, Pator, Sangot, Matebor* and *Dolloi*. The Shella confederacy was governed by *Wahadadars*. Other *himas* had *Lyngdohs*, while the institution of *Sirdar* was put in place by the British. The Garos evolved the institution of the *Nokma* of which there are four kinds, of which only the *A'king Nokma* was entrusted governance of the village. The institutions of *Nokma* and *Syiem*- particularly in the Khyrim Syiemship are unique. Whereas other tribal institutions are patriarchal in nature, these institutions are matrilineal. What we refer to as traditional institutions today must have been a matter of everyday life for the Khasi-Jaintias and Garos.

Such was the administrative, cultural and religious connection of these institutions that they remain. The operation of the Sixth Schedule of the Constitution in the Autonomous District Councils set up in 1952 have given them continued role in administration. Over the past several years these traditional institutions of governance have become increasingly active with their demand for constitutional recognition. They are finding space in contemporary governance. The State administrative machinery depends on the institutions for a great deal of support, which in turn has made the traditional heads almost indispensible within the structure of modern government. The traditional institutions have their supporters and detractors. Efforts should be made to make these institutions more relevant with the times, while still having a connection with the past.[11]

Megaliths

A striking feature of these hills is the immense number of megaliths to be found all across the highlands.[12] The Khasi-Jaintia hills have the largest concentration of such stone memorials. That the memorial stones are rough as in Nartiang and dressed such as those around Sohra/Cherrapunji may indicate that at a point in time there was the absence of the use of iron in extracting and dressing the stones while the more elaborately done up monoliths give clear signs of the use of the metal on stone. The megaliths placed in odd numbers are not gravestones. They commemorate the ancestors. Khasi oral tradition in this case becomes useful to relate when, for whom, and for what occasion were such stones erected. It is possible that the Khasi-Jaintias adopted the custom of erecting stones by the force of example or that they started its use when they finally settled in the hills. The art and ceremony was all but lost with only few stone erected in recent times. Questions have been raised how these huge stones were transported when the wheel was not used, and how they were erected, continue to test the minds of archaeologists.

Interestingly these days, Khasi-Jaintias have sought to relate more closely with this tradition. Instances of erection of memorial stones are rare. Much publicity was given to the erection of a large monolith some years ago. The Khasi Hills Autonomous District Council has erected a monolith in Khanapara, in the outskirt of Guwahati. Monoliths have become symbols of Khasi-Jaintia culture with some form of this being used by the State, Church and social organizations.

Emphasizing on three elements of the tribal culture does not preclude less importance to other facets of their life. As with tribes in the region their life was vibrant in oral traditions,[13] song, dance, and festival. The major dances are the Shad Suk Mynsiem, first organised in 1911 is annually held at springtime in Mawlai, Shillong; the Shad Nongkrem at Smit, around the Pomblang festival of Hima Khyrim,and the Wangala dance of the Garos held on a grand scale annually in some location of the Garo hills and performed in every other Garo village. Again as with other tribes in the region the Khasi-Jaintias and Garos their distinctive shawls and ornaments give the tribes identity and colour.

Influences

British rule and the several influences which that rule brought about had tremendous repercussions on the tribal societies. The incorporation of these hills and plains into the larger colonial state brought in the application of rules and regulation in their administration while retaining aspects of tribal organization in what has been called 'indirect administration'. Before the turn of the nineteenth century, the highlanders were wedged within the full force of these influences. Initial resistance as in the armed struggles of the Khasis (1829-1833), the Jaintias (1860-1862) and the Garos (1870) failed to check the British intensions. These tribal resistances recall the contribution of Tirot Sing, Kiang Nangbah and Togan Sangma to the fight against imperialism. The support many of the common folk gave to resisting imperial expansion is little remembered. Interestingly, there was no other serious challenge to British rule other than the occasional concerns of *Syiemship* succession and the disquiet Sonaram Sangma gave to the administration in the early part of the twentieth century.

The tribal societies were impacted by religion conversion by the Welsh Presbyterian mission, the American Baptist mission, the Catholic missions, and to a much less degree by the Brahmo and Vivekananda movements. They were further opened to education both by government and the Christian missions, had different occupations come their way, and were drawn into the vortex of political developments all within a carefully administered policy of tribal isolation.[14]. Resistance to the new religious influences was not met by force. Rather prominent Khasis intellectuals of the late 19th century established the Seng Khasi. Using similar methods for the propagation of their ideas as the Christian missions, the Seng Khasi has enabled Khasis to hold on to the traditional faith and preservation of culture.

Previously the interface with people not of their own tribe was limited. The fast moving changes of the 19th and 20th centuries and more so in our own times, opened the hills to interaction with traders, teachers, clerks, labourers and other job seekers from Bengal, Assam, and beyond. Many of these settlers and their descendents consider Meghalaya their home. Their contribution to the Shillong and Tura life is commendable.[15] An expression of this was the rousing reception young Amit Paul of the Indian Idol fame received when he came home. Another reflection in a sad way, of the fusion and sense of integration was expressed powerfully in Shillong when the body of Captain Clifford Nongrum, the Kargil hero came home.[16] The effect of all this has been cultural fusion and admixture. It has fostered tolerance and good will. On the other hand a section of the dominant hill society has grown intolerant to others, resulting in frequent tension. Hopefully these will become things of the past.

In More Recent Times

The close of one rule and the ushering in of freedom and self-governance brought in many more changes for the tribes and communities of Meghalaya. While their connection with Assam and its people has been long and profound, the tribal people in time asserted their right to govern themselves at higher levels of governance. The hill state movement was well underway in the late 1960s. When policy makers decided to meet the demands of the hill people, the question asked by several leaders was what should be the name of the autonomous State soon to emerge? Very appropriately the name Meghalaya was chosen. The term was first used by S.P. Chatterjee, a geographer who came to these parts in the late 1920s. He called the plateau where the Garos, Khasi and Pnars reside "Meghalaya," the adobe of clouds. First used in a doctoral dissertation entitled 'Le Plateau de Meghalaya' which was submitted to the University of Paris in 1936, the term has come to denote the state and its people. Meghalaya was made an autonomous state in 1969 (in an experiment that has not been replicated elsewhere in the country) after a peaceful demand and campaigning for a hill state.

Conclusion

Meghalaya became a full-fledged state in January 1972 following administrative reorganization of the North East. The political situation in the State has been calm if not quiet with careful adjustments over these years. Leaders in G.G. Swell and Purno Sangma emerged who contributed significantly to national politics. Many more have remained as leaders in the State from the lower rungs of the structure to their elections in the State legislature and the three Autonomous District Councils. What they will for sure do is to have a fine balance between regional and locality aspirations and fuse this within the broader spectrum of national integration. While official records speak loudly of Government's accomplishments, there are strands of disenchantment in its performance and outcome. The State has done reasonable well in terms of education, though other States have fared better; healthcare requires much attention and particularly in rural Meghalaya; agriculture remains somewhat stagnant though there is hope in horticulture and forestry; the frequent power cuts have become an annual deficiency; industrialization is a far cry; infrastructure with all its limitations in a hill state is visible and while road connectivity has definitely improved, air connectivity remains very limited.

There is much mobility between a section of people of the State who live, work and study outside the region and those who are drawn for other reasons to these hills. This is both encouraging and a matter of concern. Mobility opens up prospects that remaining within the State would never give. The State has hundreds of professions today serving in professions far from home. Entrepreneurs have emerged from the societies finding their space within the much larger opportunities offered for the enterprising. We may single out the sterling performances of the Shillong Chamber Choir which has done Meghalaya proud for its performances and awards. Meghalaya teachers, administrators, their youth and sports women and men are finding place all over the country. They have also found ways to relive and reinvent their oral traditions in song, dance and festivals.[17]

There is a refrain in a popular song: 'For The Times They Are A Changin.'[18] It applies to everyday life more than the past. We have noted the transition of the tribes of Meghalaya from their migrations, settlement, isolation and their opening and integration with the larger Indian nation, although without detail and only sketchily. It would be futile to imagine continued isolation with the fast moving time and influences. Never have tribal societies faced so many changes upon their lives and culture as they face today. Whatever happens in the years to come, the people of Meghalaya should ensure its tribal roots are cherished and retained. The uniqueness of their clan structure, matriliny and traditions should be treasured and encouraged. By maintaining their special place the people of the State will have contributed to making the world a richer, more colourful and better place.

Notes and References

*Lecture "Diversified Unity- Focus North East, The Meghalaya Perspective", Vivekananda Kendra Institute of Culture, Guwahati, 15 February 2013.

1.   S.K. Bhuyan (ed.), *Jayantia Buranji,* Gauhati,1937; S.K Bhuyan (ed.) *Deodhai Assam Buranji,* Gauhati, 1933;P.R.T.Gurdon, *The Khasis*.reproduced, N.Delhi, 1996, pp.63-65. For a more recent explanation for the migration and settlement of the Khasis refer to Glenn C. Kharkongor, "The Origin and Ancient Migration of the Khasi People: Genetics tells the Story." Paper read at the seminar 'Culture, Mission and Khasi Life', Martin Luther Christian University, Shillong, 16 February 2013.

2.   Milton S. Sagma, *History and Culture of the Garos*, Book Today, New Delhi, 1971, pp. 2-4.

3.   P.R.T.Gurdon, *The Khasis*, reproduced, N.Delhi, 1996, pp.63-65.

4.   Suniti Kumar Chatterji, *Kirata Jana Kriti,* The Asiatic Society, Calcutta, 1974, p.166; Syed Murtaza Ali, *History of Jaintia*, Dacca 1954, pp.1-6.

5.   David R. Syiemlieh, *British Administration in Meghalaya Policy and Pattern,* Heritage Pubishers, New Delhi, 1989,24-30;122-127.

6.   David R. Syiemlieh "Technology and Socio-Economic Linkages of the Khasi-Jaintia in Pre-Colonial Times", M. Momin and Cecile Mawlong (eds.), *Society and Economy in North East India,* Vol. I, Regency Publications, New Delhi, 2004.

7. J.B.Bhattacharjee, 'Brahmanical Myths, Royal Legitimation and the Jaintia State Formation', *in Social and Polity Formation in Pre-Colonial North East India* (New Delhi, 1991),p.97.; S.M.Ali, *op.cit.*, p.80.

8. Syed Murtaza Ali, op.cit., gives a fairly detailed account of Jaintiapur, the Raja's residence, its temples, the monoliths, bathing ponds, its trade connections, etc.The historical Jaintiapur is today in neglected condition.

9. Udayon Misra, *The Periphery Strikes Back: Challenges to the Nation State: Assam and Nagaland*, IIAS, Shimla, 2000, p.1-8, for a fine synoptic study of Assam and the North East.

10. Among the more detailed study of matriliny is Chie Nakane's, *Garo and Khasi: A Comparative Study in Matrilineal Systems*, Mouton & Co, Paris, 1967.

11. David R. Syiemlieh, "Traditional Institutions of Governance in the Hills of North East India: the Khasi Experience", *Man and Society, a Journal of North East Studies*, Vol. III, Indian Council of Social Science Research, North Eastern Regional Centre, Spring, 2006, pp.117-137.

12. A recent study on the Megalithic culture of the region in which there is detail of the Khasi structures is Quinbala Marak, 'Megaliths of North-East India',T. B. Subba (ed.), *North-East India: A Handbook of Anthropology*, Orient BlackSwan, 2012, pp.34-53.

13. Read the stories in Garo- with English translation in A. Playfair, *The Garos*, London, 1909 and in Khasi with English translation in P. R. T. Gurdon, *The Khasis*, reprinted Delhi, 1981.Bijoya Sawian, *Khasi Myths, Legends and Folk Tales*, Sanbun Publications New Delhi, 2010, and Esther Syiem, *The Oral Discourse in Khasi Folk Narrative,* EBH Publishers, Guwahati,2011 make interesting reading. D. S. Rungmuthu's *Folktales of the Garos*, Gauhati, 1960 could be updated.

14. Chapter III of S. K. Chaube, *Hill Politics in Northeast India,* Orient BlackSwan, New Delhi, 2012.

15. David R. Syiemlieh, 'Influence and Contribution of Bengali Settlers in Khasi Hills, Lecture delivered at the Aurobindo Ashram, Shillong.

16. Captain Keishing Clifford Nongrum, son of Peter Keishing, fought and was killed in the Kargil war in June 1999. There was spontaneous reception for this lost soldier when his body was brought back for burial. The streets of Shillong were lined for hours by mourners eagerly waiting to pay their last respect to this soldier officer.

17. Courses in traditional song, instruments and dance are encouraged at the North-Eastern Hill University, Martin Luther Christian University, St. Anthony's College and Sankardev College, Shillong.

18. Bob Dylan.

www.ingramcontent.com/pod-product-compliance
Lightning Source LLC
Chambersburg PA
CBHW021434180326
41458CB00001B/263